BURT FRANKLIN: RESEARCH & SOURCE WORKS SERIES 735

PAPERMAKING
THROUGH EIGHTEEN CENTURIES

Frontispiece: two pages of "Püchl von meim geslecht und von abentewr," the manuscript diary compiled by Ulman Stromer at Nürnberg about 1390. From the original in the German National Museum, Nürnberg

Anno lxxxxiij exeunte januario 102

Ich Ulman Stromer macht ein geding
mit Jorgen Tyrman und hab im di grossen
mül gelazzen mit den drein redern
und sol di halten von unser frawen tag
zu liechtmess für jar noch ein ander
daz erst jar hebt sich an zu liechtmess
Anno lxxxxiiij
und sol im von eder ries papirs geben
ein ort eins gul. daz sind siben ist
und sol im alle jar in den kauff machen
xxx ries gross papirs und sol daz papir
allez mit sein selbs kost auf bereiten
mit pressen und allen sachen daz ich dinne
nichtz zu schaffen haben sol dan den zeug
sol ich dar geben und ob man di redd
büzzen machen daz sol auf mein kost sein
do bey was Frantz Amman und Hanns Welser
in daz fragen Ammanns Frances

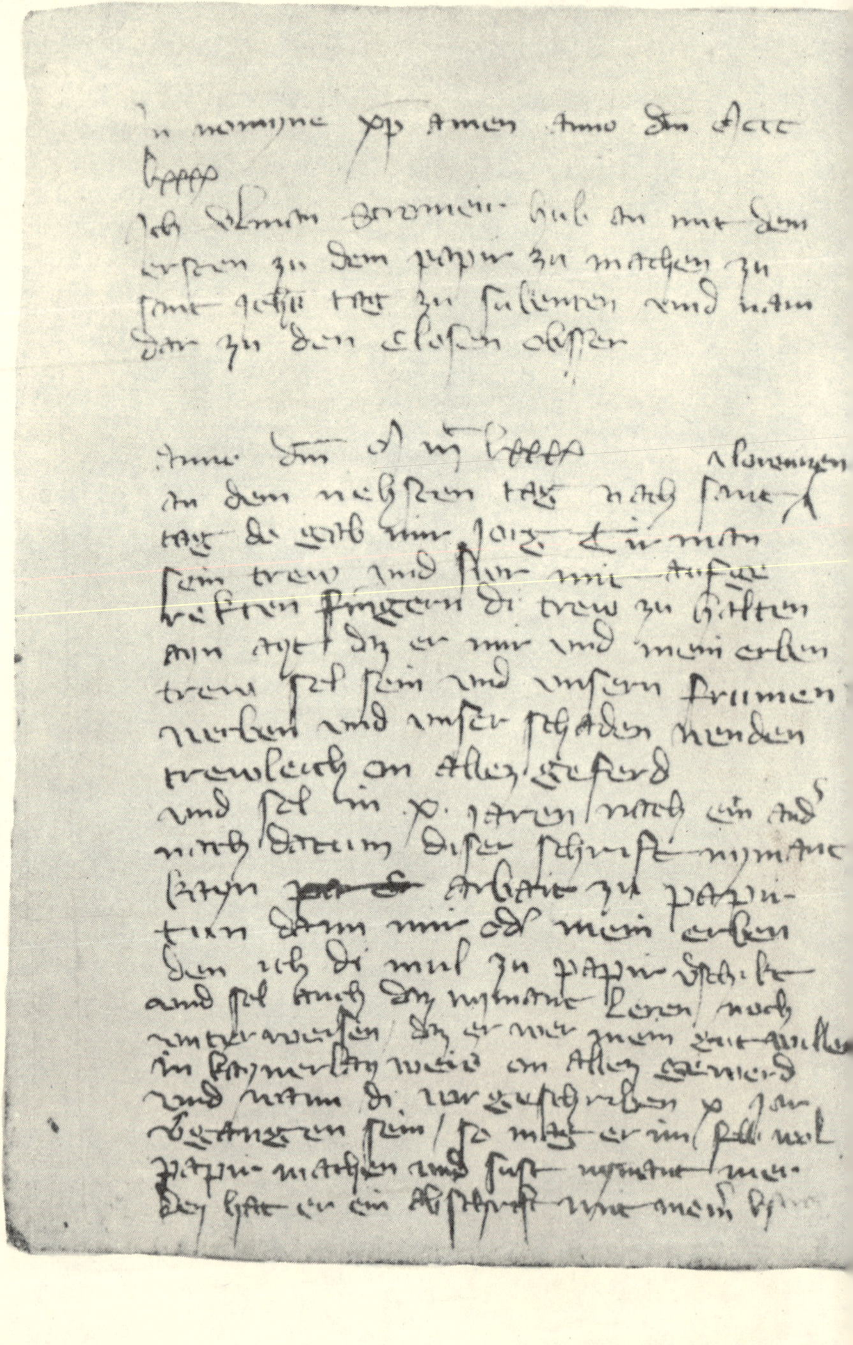
In nomine Christi amen Anno domini MCCC
XC
Ich Vlman Stromeir hub an mit dem
ersten zu dem papir zu machen zu
sant Johans tag zu sunwenden und vieng
dar zu den clesen olsster

Anno domini MCCCXC
an dem nehsten tag nach sant ...
tag do swur mir Jorg Tirman
sein trew und swur mit aufge
botten fingern di trew zu halten
ain ayt daz er mir und mein erben
trew sol sein und unsern frumen
werben und unser schaden wenden
trewleich on alles geferd
und sol in X iaren nach ein ander
nach datum diser schrift nymant
keyn ~~...~~ arbeit zu papir
machen dann mir oder mein erben
den ich di mul zu papir bestelle
und sol auch daz nymant leren noch
unterweisen daz er mer zu dem ...
in keynerley weis on alles geverd
und wenn di vorgeschriben X iar
vergangen sein so mag er im selb wol
papir machen und sust nymant mer
daz hat er ein aidschrift mit meim ...

PAPERMAKING

THROUGH EIGHTEEN CENTURIES

BY

DARD HUNTER

BURT FRANKLIN
NEW YORK

Published by LENOX HILL Pub. & Dist. Co. (Burt Franklin)
235 East 44th St., New York, N.Y. 10017
Originally Published: 1930
Reprinted: 1971
Printed in the U.S.A.

S.B.N.: 8337-17715
Library of Congress Card Catalog No.: 75-160456
Burt Franklin: Research and Source Works Series 735

PREFACE

THIS *book, while in no sense a reprint of any former work, has been compiled because of the interest shown in* OLD PAPERMAKING, *issued in limited form in* 1923. *The edition of* OLD PAPERMAKING *was limited to fewer than* 200 *copies; any more than this number, owing to the methods used in the making of the books, would have been out of the question. The demand was much greater than had been expected, and many students of printing and papermaking, as well as many bibliophiles, were unable to procure copies; the volumes now fetch prices beyond their worth. Not only was the edition of* OLD PAPERMAKING *limited in number, but owing to physical limitations in printing, the size of the book, and therefore the amount of text, were materially reduced from the original plan. The present book deals with the same phase of papermaking but in a more comprehensive manner than the older work; not only is it issued in a larger edition than was possible in the private press items, but it contains additional text and a greater variety of illustrative material. It has not, however, been possible to include original specimens, but with this particular phase of papermaking these are not essential.*

The present volume, as did OLD PAPERMAKING, *deals only with the early methods of paper fabrication and does not pretend to enter into the history of the numerous writing substances in use before the advent of paper, neither does it desire to take part in the controversy as to the priority of linen or cotton as a papermaking material; these subjects have been ably discussed in numerous works, both ancient and modern.*

A number of years ago I compiled a bibliography of the history of paper, in which hundreds of items, in all languages, were recorded. Practically all of the books and pamphlets in this list were devoted to the writing substances of the ancients, the history of papermaking materials, and the genealogy of papermaking families; a bare half dozen mention anything relating to the actual manufacture of paper in olden times; this craft has received scant attention. It is to supply this lack that the present book is issued. It aims to give the bibliophile an insight into the methods used by the old makers of paper, especially after the introduction of printing from movable types, and to interest the etcher, the engraver, and the printer, as well as those engaged in the paper and watermarking trades. With all of its omissions and shortcomings, it is the most comprehensive work that has yet been issued on this particular phase of papermaking.

DARD HUNTER

Chillicothe, Ohio
May, 1929

CONTENTS

ILLUSTRATIONS

PAPERMAKING
THROUGH EIGHTEEN CENTURIES

PAPERMAKING
THROUGH EIGHTEEN CENTURIES

PART I

THE EARLY WRITERS ON PAPERMAKING: FROM ULMAN STROMER TO MATTHIAS KOOPS

IT would be difficult to relate how I first became interested in the subject of handmade paper. In the beginning it was an interest in paper solely as a surface upon which to print, later I became engrossed in the actual forming of the sheets; living in America where none of the work was being done, I had no way of ascertaining at first hand just how paper of this kind was fabricated. There were few books available in the libraries, and apparently the subject was one that had been very much neglected. I knew how paper was made in a continuous roll on a machine, but this new method did not interest me, nor have I since been inquisitive regarding this mode of producing paper. It seems only to be a matter of how wide the machines can be made and how many feet of paper per minute they will expel. So in reality I am not interested in just *Paper,* it must be either the material that is moulded in the ancient Oriental or European manner, or the primitive paper that is beaten into sheets by hand in the outlying corners of the world. The improved paper machine is essential to cope with modern life as it has come to be, but I would be content never to see again one of those long, ponderous, steaming dragon-like machines emitting from its great jaws the streams of paper that are thought necessary to our very existence. Inasmuch as I am a user, indirectly, of several hundred tons of rolled paper a year, I trust my good friends, the papermakers will not censure me too severely for speaking so frankly and unkindly regarding the modern version of the invention of Mr. Nicolas-Louis Robert.

After working for a number of years with European and Oriental handmade papers and developing a genuine affection for them, it was my desire to know exactly how this type of paper was made, and to discover what constituted the elusive features that fascinated me to such an extent, so I journeyed to Europe with the express purpose of acquiring a knowledge of the craft. This was about twenty-five years ago, and it will not be necessary to enter into a relation of the myriad difficulties encountered. Most of the European masters were exceedingly considerate, and I am profoundly grateful to them for their many kindnesses, both at the time of my first visit and through the years that have followed. It was through the courtesies of the English and continental makers of handmade paper that my initial knowledge of the art as it is practiced in modern times was acquired.

With a little knowledge of the making of handmade paper in modern times came a desire to know how paper had been fabricated in the past, and to learn something of the traditions and history of this ancient and noble craft. There was, at that time, little or no interest in the subject of papermaking, aside from the usual routine of modern equipment, production, and selling,—the historical and romantic side of this handicraft, extending over a period of almost two thousand years, had apparently not caught the imagination of those engaged in the making of paper, either by hand or machine. It was surprising to find the libraries of the wealthy mill-owners, in both Europe and America, absolutely devoid of bibliographical material relating to the history of the industry to which they were devoting their lives. I knew that there must be many books and pamphlets on the subject of old watermarks and papermaking, as I had myself for years, in a modest way, been collecting volumes relating to printing and type founding, subjects that had not been neglected, many ardent and persistent collectors gathering in every available old typographical book and type specimen sheet. On the subject of papermaking there were, of course, such books as Herring's and Munsell's to be found on the shelves of every second-hand bookseller, for, while excellent works, they had been printed in large editions and were

NO. I

Ulman Stromer, the first to write about papermaking

not rare in any sense. The old and little-known books were the ones that I most desired and after a year's search in the book-shops of London and on the continent, I managed to find about thirty volumes dealing with the subject of old papermaking, and these few books formed the nucleus of my most incomplete collection. For a few shillings or marks items have been purchased which are most highly prized and which would not be parted with for many times their original cost. Old books on papermaking and watermarking will never be much sought after, but even so the prices mount higher each year. These items, however, are not for the most part the excessively rare or unique volumes such as would insnare such profound bibliophiles as Dr. A. Edward Newton or Dr. Rosenbach; but, nevertheless, many of the things are exceedingly difficult to find, and years have been spent in the acquisition of certain obscure books and pamphlets. It is fortunate that this hobby does embrace a field where money counts not nearly so much as do time, patience, and persistence. Books that interest me from a collector's standpoint seldom find their way to the auction room, one must seek them out from remote and hidden places, as there are few bibliophiles in this particular class of material, and its collecting receives no publicity,—"paperiana" seldom comes up at the "great sales."

The first bibliography that I compiled was completed in 1915 and listed only the books that related directly to the watermarks of handmade papers. The compilation embraced 167 items; now half as many more titles could be added to the list. A collection is never complete; no matter how large it may become there is always something more. I now have all of the items listed in this bibliography. Some are pamphlets of only a few pages; other titles are mere magazine articles, scraps gathered here and there in many languages, any one of which would not cost so much as some first editions of modern fiction, but the time and patience required in their assembling were considerable. In the more modern works it is the thin, obscure, paper-covered pamphlets that are difficult to locate, as these were printed in small editions, and after being once read were cast aside as of no future use or value; the big, bound volumes turn up readily, but the leaflets and brochures on papermaking and watermarking are elusive.

To anyone interested in gathering together a collection of

"paperiana" I would first suggest that all the bibliographies be studied and lists of the desired items compiled. Then communicate with all of the antiquarian booksellers of Europe and America, no matter how remote or insignificant they may be; once they

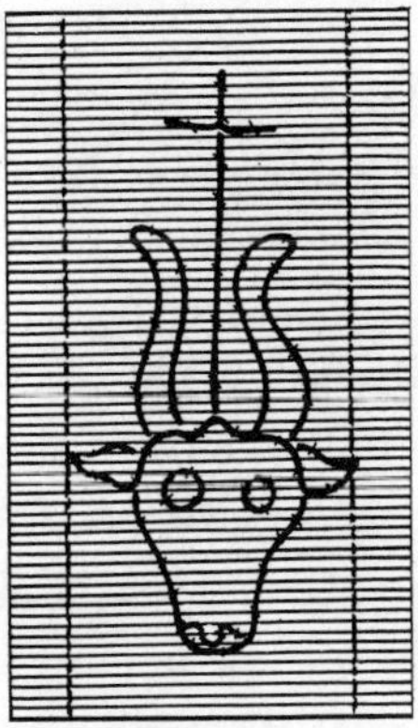

NO. 2

The bull's head watermark appears in the paper on which Ulman Stromer wrote his diary

learn that you are interested in a certain subject you will receive "reports" from time to time,—more than likely regarding items that you already possess. A collection of this nature is formed over a period of many years, the items cannot be amassed hurriedly, no matter how much money one may have at his disposal.

Aside from the collecting of books and prints, there are old and modern watermarked papers to be gathered together, as well as decorated papers such as were used in years past for the end-papers of books; there are also the unprinted money- and passport-papers of the various governments. The pursuit is unlimited as there are literally thousands upon thousands of these papers which could be assembled and catalogued, to say nothing of old manuscript letters, wills, deeds, and the like, which contain watermarks in the paper upon which they were written; these, too, are inexhaustible; also labels, proclamations, maps, and charts of the old papermakers. Then there are the native Oriental handmade papers, both old and modern, from China, Japan, Indo-China, and India waiting to be collected, as well as specimens of the barks, grasses, and plants from which they were made, and there are also the hand-beaten mulberry-bark papers of the South Sea Islands, and the more crudely made bark-papers of South America and Central Africa. Should one become a real enthusiast it is interesting to gather the various types of moulds which have been employed in ages past for the actual forming of sheets of paper throughout the world, also the tools and appliances of the "tapa" makers of the Pacific Islands and other strange sections of the earth. The deeper one gets into this mania for assembling "paperiana," the larger the scope seems to become; and the collector, after a number of years of experience, begins to realise that a complete assemblage of this particular kind can never be a reality, either as a public or private endeavour. In other branches of bibliomania it may be entirely feasible to acquire all of the first editions of an esteemed author, or a complete group of authors, or even all the works on a less complicated subject, but no one will ever be able to say that he has gathered under one roof *all* of the multitudinous material relating to the apparently limited subject of Paper.

In giving an account of the most important books in this category it will not be possible to venture beyond the year 1800, and only Occidental works will be noted. The greatest and most prized of all works relating to this subject is a manuscript by Ulman Stromer in the German National Museum in Nürnberg, and no matter how ardent or enthusiastic the collector may become, he must be content with a photographic facsimile of this rare volume, for it will probably always, as it rightly should, remain in the city of its origin. It would be hard to speculate as to what this diary of a few written pages would fetch at auction, but I venture to say that it would lead the Gutenberg 42-line Bible a good race.

Ulman Stromer was the first writer to touch on the art of papermaking, and he compiled a diary which dealt with his tribulations

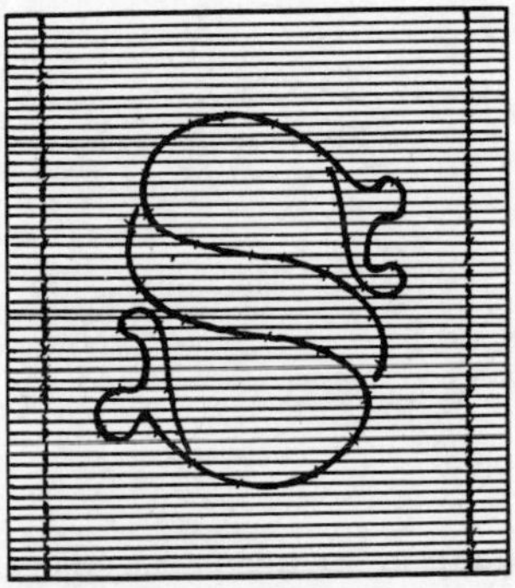

NO. 3

The "S" watermark which is thought to have been used by the Stromer mill. This example is from the year 1397

and hardships in founding a paper mill. This manuscript was entitled, "Püchl von meim geslecht und von abentewr." It was begun about 1390 and is entirely in the calligraphy of the author. Stromer was a member of the senate that governed Nürnberg and his mill was located just outside the western wall of the city, at the point where the Pegnitz River emerged into the fields. Before the papermaking venture Stromer had been a merchant, and on his journeys of trade to northern Italy he had seen paper mills in operation. He was convinced that he would be successful with an establishment of the same kind in his own country, especially if he could secure a monopoly in the work. No one in Nürnberg knew anything about making paper, so it was necessary that the erstwhile trader procure his workers elsewhere. He set out to find several foreigners who were skilled in the art of papermaking, and after considerable deliberation Stromer succeeded in inducing a number of artisans to leave their positions in Lombardy.

Ulman Stromer was, at this time, sixty-two years of age, and his experience in trading had made him keen and of a cautious dispo-

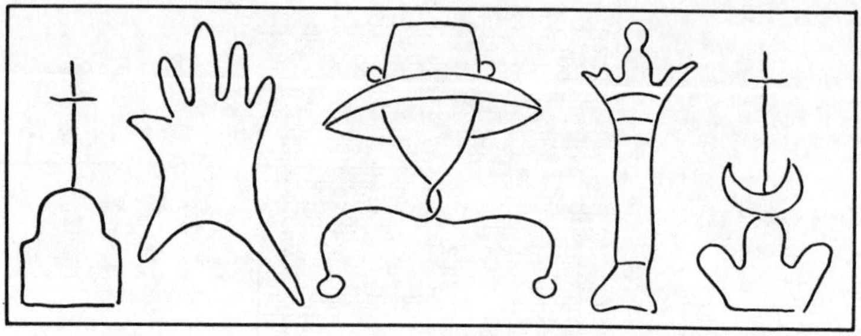

NO. 4

These watermarks appear in the paper upon which the Ripoli press cost-book was written. A reproduction of a page from this manuscript is shown in illustration No. 147

sition. After procuring the workers he had all of them swear to be loyal to him and keep the secrets of their craft to themselves so that there would be no danger of other paper mills springing up in Germany. Accordingly, Ulman Stromer wrote in his manuscript diary that Closen Obsser, the man he had engaged as foreman, "promised to be faithful unto me and declared on his oath that he would be true to me and my heirs, that he would be my overseer at the mill, keep me from harm and that as long as he lived he would make paper for no one else save me and for my heirs and would not teach any man to make paper in any way at all. This took place on the Sunday next before St. Lawrence Day, in my room at the time of evening prayer in the year 1390 when my son Jörg was present." Thus having his foreman's faithfulness assured, Stromer proceeded to have the rest of his employees take the same oath. He recorded in his chronicle that, "it was on the day after St. Lawrence when Jörg Tyrmann swore on oath to the Saints that he would help my progress materially and that for ten years he would make paper only for me and my heirs and would not instruct anyone except at my sanction. But when the ten years are passed he may make paper for himself, but for no one else and he may then teach those who make paper for him, but for no one else, as long as he lives."

These men, like himself, were Germans, but with the skilled Italian artisans Stromer felt that an extra precaution was essential, so he took them before a procurator named Conradus, and Stromer's sons, brother, and brother-in-law acted as witnesses to the affidavits. The contract with the Italian workmen read: "In the year 1390, Franciscus de Marchia, and Marcus, his brother, and his manservant Bartholomeus pledged their loyalty to me and swore on oath to the Holy Saints that they would forever be faithful and would not divulge the secrets of papermaking to anyone in all the German lands this side of the mountains of Normandy."

The precautions taken by Stromer to guard his own interests rather excited the suspicions of the Italians, they felt that if they were so indispensable to the success of the paper mill it would be better for them to control the enterprise than for Stromer to be the master. With this idea in view the Lombardians tried to hamper the paper mill on every side, hoping that Stromer would become weary and discouraged with the whole venture and lease the mill to them at a low figure so that they might operate it as they liked. Continuing with his narrative Stromer wrote, "the Italians were most troublesome the first year and did as they pleased and caused me many difficulties, not suffering the third wheel to run at all. My first two wheels ran eighteen stampers and even these were left idle a great part of the time as the foreign workmen wished to produce as little paper as possible so as to force me to let them have the mill for a rent of 200 gulden a year. This I would not agree to do and they then offered also to give me an amount of paper. From this I understood that they wished to ruin me and deprive me of my paper mill."

In mediaeval times the master had great authority over his workers and Stromer recorded in his diary: "In the year 1391, on the twentieth of August, I took Franz and his brother Marcus and shut them in the tower." On the fourth day of their imprisonment the unruly Italians sent for Stromer's brother-in-law, thinking that he would be more lenient with them than their master. A reconciliation was effected and the men were liberated on the fourth day, but not until they had sworn that they would cause no further trouble and that they would thereafter do their full share of work at the paper mill.

The confinement in the tower had the desired result upon the scheming Italians and they worked more in harmony thereafter. Encouraged by the improved discipline Stromer engaged additional papermakers, who were duly sworn like the others. A carpenter was needed to repair the stampers and vats, for, in 1392, Stromer recorded: "In my back room Erhart Zymerman has been engaged to be my servant for one year. He is to do the carpentry work at the mill or he will polish paper, and his wife will sort rags, or hang paper on lines to dry, or count the paper. Wages will be paid them, a good house in which to live and firewood withal." Ulman Stromer operated the paper mill from 1390 to 1394 when he leased the property to Jörg Tyrmann, for a period of four years.

The paper on which the Stromer manuscript was written reveals the bull's head and cross watermark in three different forms, but it is thought that the paper was not from the Stromer mill but of Italian fabrication. The "chain-lines" of the paper are widely spaced, and the "laid-lines" heavy and coarse. The initial "S" was used on some of the Ulman Stromer moulds as a watermark, but in some of the paper that was made in this fourteenth century mill the arms of Nürnberg appeared as the device. Stromer died in the year 1407 at the age of seventy-nine. This manuscript diary constitutes the most valuable document in the annals of papermaking.

In chronological order, but not in papermaking importance, the next volume to be listed would be the manuscript cost-book of the Ripoli press, which was compiled in Florence from 1474 to 1483. This manuscript (13 x 19 cm.) is now in the Magliabechian Library, Florence, where it should always remain. The book has a papermaking interest inasmuch as it gives the names and prices of paper in use for printing at that time. These names include, "large paper of Bologna in common folio, paper of Fabriano with cross-bow for watermark, the same with cross watermark, paper of Colle, paper of Prado, paper of Pescia with spectacles for watermark, the same with glove or hand for watermark." The prices would work out at about six dollars for a ream of folio paper. This book, like that written by Ulman Stromer almost a hundred years earlier, is beyond the collector, but photographic copies may be had of both by applying to the directors of the museum or library where they are deposited. There is one item, however, which the paper bibliophile should acquire, and that is a copy of Schedel's *Liber Chronicarum,* which was printed in Nürnberg by Anthony Koburger in 1493. In this book, familiarly known as the "Nürnberg Chronicle," there is a double page woodcut of the city of Nürnberg and in the lower right corner of this illustration a group of buildings is shown. These half-timbered structures are supposed to have been the former paper mill of Ulman Stromer.

The next volume of papermaking interest is *Bartholomæus: De Proprietatibus Rerum* printed by Wynken de Worde, the successor of William Caxton, at Westminster in 1498. This book is considered to be the first book printed in England in which English-made paper was used. The paper was fabricated by John Tate, who

Sexta etas mūdi
NVREMBERGA

NO. 5

The engraving of Nürnberg from Schedel's *Liber Chronicarum* of 1493. The group of five buildings, outside the wall, was Ulman Stromer's paper mill

operated a mill in Hertford, the watermark being in the form of an eight-pointed star enclosed in a double circle, resembling a cart wheel. John Tate was a mercer in London, and the son of John Tate who was Lord-mayor in 1473. Tate, the younger, died in

NO. 6
The arms of Nürnberg watermarks

1507, his will containing several references to the paper mill. In the first place he bequeathed to Thomas Bolls of Hertford, "as moche whit paper or other paper as shal extende to the summe of xxvi s. viij d . . . owte of my paper mill at Hertford." Then he directed his executors to dispose of the paper mill, "with all the goods, woods, pastures, medes, with all the commodities concerning the said myll to the moost advantage." Finally, in leaving to his eldest son Robert, all his lands in Hertfordshire and Essex, the paper mill was excluded: "My paper myll with the appurtenaunces to be always excepted and to be sold." The Tate paper mill was evidently short lived, as there is no evidence of its existence before 1495 or after 1498. The following lines appear in the Wynken de Worde English edition of *Bartholomæus: De proprietatibus Rerum* showing without doubt that the paper used in this book was made in the John Tate mill:—

And John Tate the yonger Ioye mote he broke
Whiche late hathe in Englonde doo make this paper thynne
That now in our englyssh this boke is prynted Inne

NO. 7

The watermark of John Tate, the first English papermaker

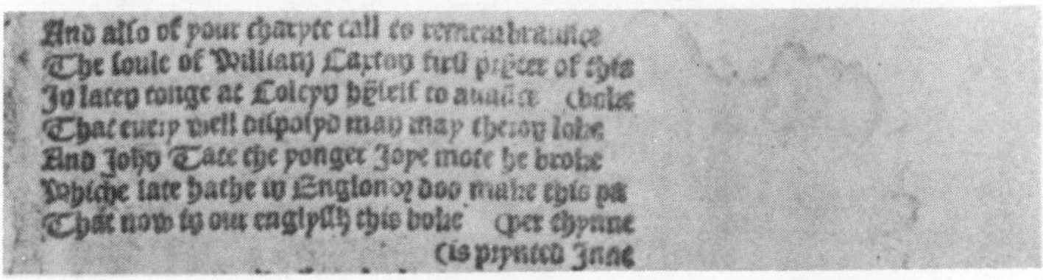

And also of your charyte call to remembraunce
The soule of William Caxton first pryter of this
In laten tonge at Coleyn hymself to auaunce boke
That every well disposyd man may theron loke
And John Tate the yonger Joye mote he broke
Whiche late hathe in Englond doo make this pa
That now in our englysshe this boke per thynne
is prynted Inne

NO. 8

Paper from the John Tate mill was used in printing the Wynken de Worde edition of *De Proprietatibus Rerum* in 1498. This was the first use of English-made paper in book printing

In 1568 there was published in Frankfort-on-Main a little book of trades in which are pictured one hundred and fifteen trades or callings, each depicted by that remarkable wood-engraver, Jost Amman (1539–1591); under each engraving there is a verse about the particular calling or craft, written by Hans Sachs (1494–1576),

NO. 9

Jost Amman, the first engraver to depict the craft of papermaking

the cobbler-poet of Nürnberg. The woodcut of the papermaker is the eighteenth in the book and follows the illustration picturing the woodcutter at work. This is the first delineation of the papermaker made and it shows the vatman forming a sheet of paper, a

Der Papierer.

Ich brauch Hadern zu meiner Mül/
Dran treibt mirs Rad deß Wassers viel/
Daß mir die zschnittn Hadern nelt/
Das Zeug wirt in Wasser eynquellt/
Drauß mach ich Bogn/auff den Filtz bring/
Durch Presß das Wasser darauß zwing.
Denn henck ichs auff/laß drucken wern/
Schneeweiß vnd glatt/so hat mans gern.

J ij Der

NO. 10

The engraving by Jost Amman and the poem by Hans Sachs in the German book of trades of 1568

NO. 11

Hans Sachs, the cobbler-poet of Nürnberg, the first writer to compile a poem about papermaking

Eygentliche Beschreibung
Aller Stände auff Er-
den/ Hoher vnd Nidriger/ Geistlicher
vnd Weltlicher/ Aller Künsten/ Handwercken
vnd Händeln/ ꝛc. vom grösten biß zum kleinesten/
Auch von jrem Vrsprung/ Erfindung vnd
gebreuchen.

Durch den weitberümpten Hans Sachsen

Gantz fleissig beschrieben/ vnd in Teutsche Reimen ge-
fasset/ Sehr nutzbarlich vnd lustig zu lesen/ vnd auch mit künstreichen
Figuren/ deren gleichen zuvor niemands gesehen/ allen Ständen
so in diesem Buch begriffen/ zu ehren vnd wolgefallen/ Allen
Künstlern aber/ als Malern/ Goldschmiden/ ꝛc.
zu sonderlichem dienst in Druck
verfertigt.

Mit Röm. Keys. Maiest. Freyheit.
Gedruckt zu Franckfurt am Mayn.
M. D. LXVIII.

NO. 12

Title-page of the German book of trades, in which appear the engraving by Jost Amman, and the poem by Hans Sachs relating to papermaking

small boy carrying finished paper, the press and the stampers. The title of this sixteenth century book of trades is: *Eygentliche Beschreibung Aller Stände auff Erden, hoher und nidriger, geistlicher und weltlicher, aller Künsten, Handwercken und Händeln.* A translation of the verse by Hans Sachs underneath the woodcut of the papermaker, reads:

Rags are brought unto my mill
Where much water turns the wheel,
They are cut and torn and shredded,
To the pulp is water added;
Then the sheets 'twixt felts must lie
While I wring them in my press.
Lastly, hang them up to dry
Snow-white in glossy loveliness.

An early book to treat of technical papermaking was compiled by Samuel Zimmerman and issued in 1579.[1] While there is nothing in this volume of any great importance regarding sixteenth century papermaking, it is interesting inasmuch as it mentions the manufacturing of paper and its various qualities and uses. It also advances, for the first time, the use of paper as a substitute for leather in bookbinding. Another book of papermaking interest that appeared in the sixteenth century was the volume by Tommaso Garzoni (1549?–1589),[2] printed in Venice in 1586. The short article that deals with paper is insignificant and is interesting only on account of the early period represented.

Of the old papermakers of England, John Spilman is today the best known, this being due to a poem which was written by

1 New Titularbuech Das ist Wie man ainer Jeden Person sey was Würdigkait sie wölle Bayde des Gaistlichen und Weltlichen Stands zuschreiben soll. . . . Sambt etlichen hinzugethanen fürtreffentlichen Gehaimnussen verborgenen Mechanischen Apocryphischen und gleichsam ubernatürlichen Künsten das Lesen und die Schreiberey betreffendt Dergleichen vor niemals in Truck aussgangen. . . . Ingolstadt, D. Sartorius, 1579.

2 La piazza vniversale di tvtte le professioni del mondo, e nobili et ignobill nvovamente formata, e posta in luce de Tomaso Garzoni . . . Venice, 1586.

Thomas Churchyard (1520?–1604) in 1588, and also to the fact that it is thought Spilman's mill was referred to by Shakespeare in *King Henry VI.* Spilman was a German, a native of the town of Lindau on Lake Constance, and after moving to England be-

NO. 13

Samuel Zimmerman the earliest writer to suggest the use of paper for bookbinding and other bibliographical purposes

came one of Queen Elizabeth's goldsmiths. He leased a mill on the Darent in the royal manor of Bicknores at Dartford, Kent, which he converted into a paper mill, known to have been working in 1588. Early in 1589 he obtained a patent which gave him a monopoly of papermaking and the collecting of rags in the Kingdom. After holding this patent for eight years he surrendered it, and in its place another covering a period of fourteen years was given to his mill. This embraced the making of white writing paper and paper of all other descriptions. It seems that there

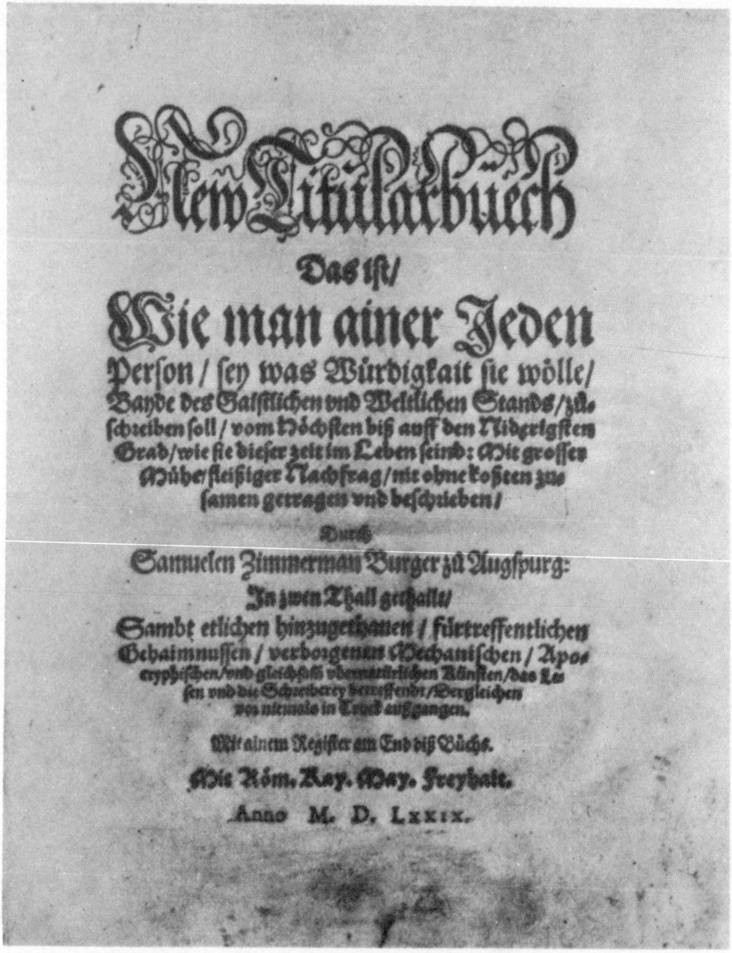
New Titularbuech
Das ist/
Wie man ainer Jeden
Person / sey was Würdigkait sie wölle/
Bayde des Gaistlichen vnd Weltlichen Stands/zu-
schreiben soll / vom Höchsten biß auff den Nidrigsten
Grad/wie sie dieser zeit im Leben seind: Mit grosser
Mühe/fleißiger Nachfrag/nit ohne kosten zu-
samen getragen vnd beschrieben/
Durch
Samuelen Zimmerman Burger zu Augspurg:
In zwen Thail gethailt/
Sambt etlichen hinzugethanen / fürtreffentlichen
Gehaimnussen / verborgenen Mechanischen / Apo-
cryphischen/vnd gleichsam vbernatürlichen Künsten/das Le-
sen vnd die Schreiberey betreffendt/Dergleichen
vor niemals in Truck außgangen.
Mit ainem Register am End diß Büchs.
Mit Röm. Kay. May. Freyhait.
Anno M. D. Lxxix.

NO. 14
The title-page of the book by Samuel Zimmerman

were several infringements of Spilman's privileges, but they were evidently settled peacefully, as the papermaker-goldsmith was knighted in 1605, upon a visit paid by King James to the paper mill at Dartford.

LA

PIAZZA

VNIVERSALE

DI TVTTE LE PROFESSIONI
DEL MONDO, E NOBILI
ET IGNOBILI.

*NVOVAMENTE FORMATA, E
posta in luce da Tomaso Garzoni
da Bagnacauallo.*

AL SERENISS.MO ET INVITTISS.MO
ALFONSO SECONDO DA ESTE
DVCA DI FERRARA.

CON PRIVILEGIO.

IN VENETIA,
Appresso Gio. Battista Somascho.

M D L X X X V I.

NO. 15
The title-page of the volume by Tommaso Garzoni

NO. 16

Sir John Spilman's tomb in the Dartford Parish Church, Dartford, Kent. The effigies are coloured and are of Sir John, who died November 1626, and his first wife, Lady Elizabeth, who died May 1607. The monument was restored and removed to its present situation before 1876 by the Fraternity of Papermakers

Shakespeare's reference to a paper mill occurs in *King Henry VI,* in a speech of Jack Cade to Lord Say: "Thou hast most traitorously corrupted the youth of the realm in erecting a grammar-school; and whereas, before, our fore-fathers had no other books but the score and the tally, thou hast caused printing to be used; and, contrary to the king, his crown, and dignity, thou hast built a paper-mill."

This play was written in 1623, three years before Spilman's death, and it may be that Shakespeare had this mill and its monopoly in mind when he wrote the lines. Jack Cade's rebellion was in 1450, and there can be no doubt that the Lord Say had nothing whatever to do with paper mills. His son, however, acquired by marriage a property at North Newton, Oxfordshire, and it is known that North Newton Mill was at one time a paper mill; this has led some writers to take Shakespeare's words literally and to advance them as evidence establishing this North Newton Mill as the first paper mill in England.

The poem by Thomas Churchyard referred to regarding the mill of John Spilman, first appeared in 1588, under the title of: *A Discription and playne discourse of paper, and the whole benefites that Paper brings, with rehersall, and setting foorth in verse a Paper myll built nere Darthford, by a High Germaine, called Master Spilman. Ieweller to the Queenes Maiestie.* The complete poem consists of 353 lines printed in a quarto volume of fifteen pages, with bordered title-page. While this description of paper is of no value as a technical history, it is, nevertheless, of great interest as the first description of papermaking to appear in English. The work was prefaced by a sort of dedicatory epistle addressed, "To my honorable friend Sir Walter Raleigh," under the title: "A Sparke of Frendship and Warme Goodwill, etc." The composition is much too long to reprint in its entirety, a few salient verses that deal particularly with the subject of paper will suffice:

A SPARKE OF FRENDSHIP AND WARME GOODWILL, THAT SHEWES THE EFFECT OF TRVE AFFECTION AND VNfoldes the finenesse of this world.

VVhereunto is ioined, the commoditie of sundrie Sciences, the benefit that paper bringeth, with many rare matters rehearsed in the same: with a description & commendation of a Paper Mill, now and of late set vp (neere the Towne of Darthford) by an high Germayn called M. Spelman, Jeweller to the Queenes most excellent Maiestie.

Written by THOMAS CHVRCHYARD Gent.

Nulla potest esse iucunditas, sublata amicitia.
Cic. pro Flax.

Printed at London. 1588.

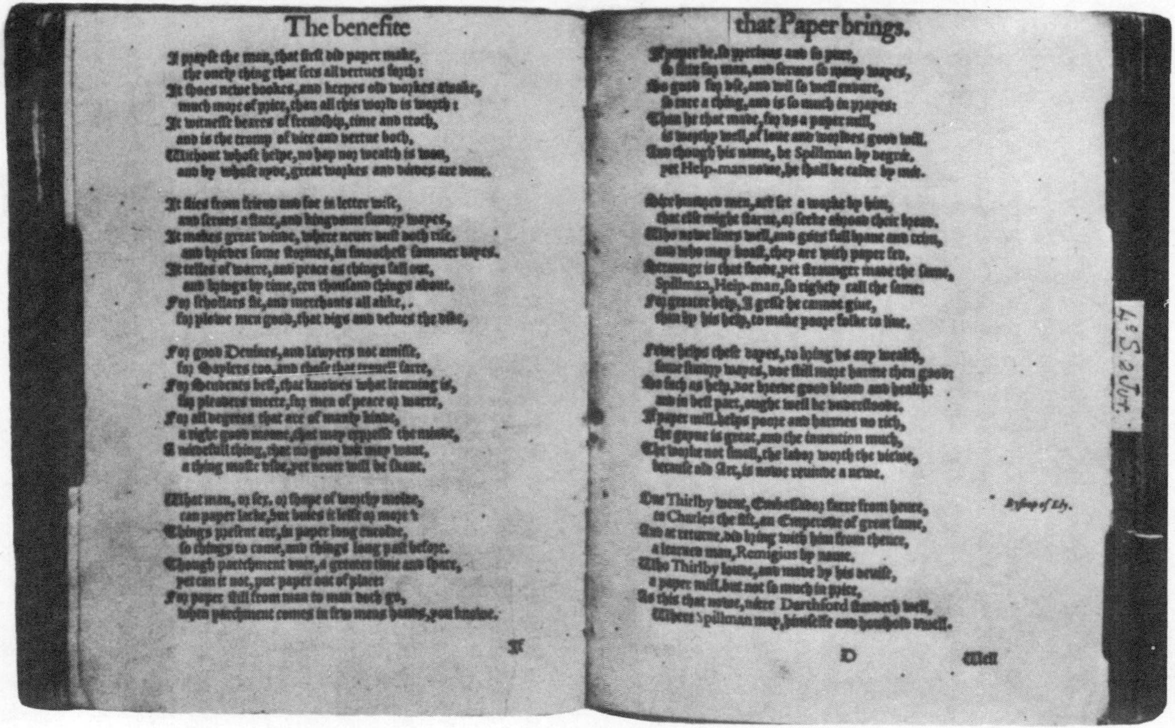

NO. 17

The title-page and first eight verses of the poem by Thomas Churchyard regarding the Sir John Spilman paper mill

I prayse the man, that first did Paper make,
the onely thing that sets all vertues forth:
It shoes new bookes, and keeps old workes awake,
much more of price than all this world is worth;
Though partchment duer, a greater time and space,
yet can it not, put paper out of place:
For paper still from man to man doth go,
when partchment comes in few mens hands, you knowe.

Then he that made, for us a Paper mill,
is worthy well, of love and worldes good will,
And though his name, be Spillman by degree,
yet Help-man nowe, he shall be calde by mee.
Six hundred men, are set a worke by him,
that else might starve, or seeke abroad their bread,
Who nowe lives well, and goes full brave and trim,
and who may boast they are with paper fed.

The hammers thump, and make as lowde a noyse,
as Fuller doth, that beates his woolen cloth,
In open shewe, then sundry secrete toyes,
makes rotten ragges, to yeelde a thickned froth:
Then is it stampt, and washt as white as snowe,
then flong on frame, and hang'd to dry I trow:
Thus paper streight, it is to write upon,
as it were rubde, and smoothde with slicking stone.

We are indebted to Vittorio Zonca for the first delineation of a stamping-mill in which the machine was shown in detail. This appeared in 1607 in an Italian book of machinery with the title: "Novo teatro di machine et edificii per uarie et sicure operationi," published in Padua. A reproduction of the Zonca engraving is given in Part III. Then came, grouped here with Zonca's book merely for convenience, and because of a certain chronological order, the works of Guidonis Pancirolli, (*Rerum Memorabilium sive Deperditarum, pars prima et secunda,* Francofurtens, 1660),

NOVO TEATRO
DI MACHINE ET EDIFICII
Per uarie et sicure operationi
Cō le loro figure tagliate in Rame
e' la dichiaratione, e dimostratione
di ciascuna.
Opera necessaria ad Architetti, et a quelli
ch. di tale studio si dilettano.
DI
VITTORIO ZONCA Architetto
della Magnifica Communità
di Padoua.
Al Ill.mo Sig.or Prone mio Col.mo Il Sig.r
RANVCIO Gambara Conte di Virola Algisi
et c.
In Padoua appresso Pietro Bertelli.
1607

NO. 18

The title-page of the volume on machinery by Vittorio Zonca. In this book appears the first illustration of a set of stampers for the maceration of papermaking material and the Spanish author Pedro Mexía (1496–1552), (*Silva De Varia Lección en la qual se tratan muchas cosas muy agradables y curiosas,* Madris, 1662). In the first of these two works papermaking was treated in an insignificant way on pages 251–257, in the second part of the volume; while in the work by Mexía pages 306–310 are devoted to the origin of paper. It is curious that, although papermaking had its European origin in Spain, there are very few Spanish books, either ancient or modern, that deal with the craft.

We now come to the important book of Georg Andreae Böckler which was printed in Nürnberg in the year 1662 under the title of, *Theatrum machinarum novum exhibens opera molaria et aquatica constructum industria.* In this volume there are 154 full-page plates depicting various types of machinery, each engraving having a short explanation in Latin. Two plates are devoted to papermaking and present the clearest delineation of the art to this date. One plate shows the water wheel, stampers, vat, press, and the finished paper hung to dry. The vatman is in the act of forming a sheet of paper, and another workman is operating the press, while bundles of rags appear in the foreground. The other plate gives a clear conception of the stampers which were used for the maceration of rags at the inception of papermaking in Europe. These engravings are reproduced in Nos. 74 and 107. A translation of the Latin description of this plate reads: "On the river bank stands conveniently the water-mill wheel which turns an axle fitted with cams. These various cams in turn actuate the stamps, lifting them up and letting them fall full force upon the rags. Through the trough water from the river is distributed to the receptacle which serves to hold the rags." It is quite evident that the engraver who made the plate for Böckler used the earlier picture from the volume of Vittorio Zonca for the model.

In the *History of the Worthies of England,* Thomas Fuller, (1608–1661), gives some observations regarding the paper of his time. The book was published in London in 1662 and is arranged by counties, with the commodities, manufactures, etc. of each, followed by brief biographies of its "worthies." He says of paper: "Paper participates in some sort of the characters of the country which made it; the Venetian being neat, subtle, and court-like; the French, light, slight, and slender; and the Dutch, thick, corpulent, and gross, sucking up the ink with the sponginess thereof. To such who object that we can never equal the perfection of Venice paper, I return, neither can we match the purity of Venice glasses; and yet many green ones are blown in Sussex, profitable to the makers, and convenient to the users. Our home-spun paper might be found beneficial."

The earliest reference in any publication to the Rittenhouse mill or to papermaking in America appeared in Philadelphia in 1692 and was written by Richard Frame, the title of the book being: *A short description of Pennsilvania, Or, A Relation of What things are known, enjoyed, and like to be discovered in the said Province. Presented as a Token of Good Will to the people of England.* The Rittenhouse mill, as noted in Parts V and VII, was the first in the colonies, having been established near Germantown, Pennsylvania, about 1690. The verses are printed in full:

> The German-Town, of which I spoke before,
> Which is, at least in length one Mile or more,
> Where lives High-German People, and Low-Dutch,
> Whose Trade in weaving Linen Cloth is much.
> There grows the Flax, as also you may know,
> That from the same they do divide the Tow;
> Their trade fits well within this Habitation,
> We find Convenience for their Occasion.
> One Trade brings in employment for another,
> So that we may suppose each Trade a Brother;
> From Linen Rags good Paper doth derive,
> The first Trade keeps the second Trade alive;

Without the first, the second cannot be,
Therefore since these two can so well agree,
Convenience doth appear to place them nigh,
One in German-Town, t'other hard by.
A Paper-Mill near German-Town doth stand.
So that the Flax, which first springs from the Land,
First Flax, then Yarn, and then they must begin
To weave the same, which they took pains to spin.
Also, when on our backs it is well worn,
Some of the same remains Ragged and Torn;
Then of the Rags our Paper it is made,
Which in process of time dost waste and fade:
So, what comes from the Earth, appeareth plain,
The same in Time, returns to Earth again.

It was not until 1696 that the first mention of the actual fabrication of paper was made in English. This appeared in a book of machinery by Venturi Mandey, published in London under the title, *Mechanick Powers; or, The Mystery of Nature and Art unvail'd. Shewing what Great Things may be perform'd by Mechanick Engines, in removing and raising Bodies of vast Weights: with little Strength or Force; also the making of Machines or Engines, for the raising of Water, draining of Grounds, and several other Uses.* This volume gives the first description in English of the appliances that were used in the maceration of papermaking material. One reference (page 71) reads: " . . . Also Engines to make Paper, in which a wooden Cylinder of sufficient magnitude standing out, furnisht with little troughs, takes up wooden Mallots, which bruise and break whatever is laid under them to powder."

The following verse was written by John Holme in 1696 under the title of *A True Relation of the Flourishing State of Pennsylvania* and is usually considered to have been the earliest metrical composition composed in Pennsylvania, but it was not published

Mechanick Powers:

OR, THE

MYSTERY

OF

NATURE *and* ART

UNVAIL'D

Shewing what *Great Things* may be perform'd by Mechanick Engines, in removing and raising Bodies of vast Weights: with little Strength or Force; and also the making of Machines or Engines, for raising of Water, draining of Grounds, and several other Uses.

TOGETHER,

With a Treatise of *Circular Motion* artificially fitted to Mechanick use, and the making of Clock-work, and other Engines.

A Work pleasant and profitable for all sorts of Men, from the highest to the lowest Degree: And never treated of in *English* but once before, and that but briefly.

The whole comprized in Ten Books, *and illustrated with Copper Cuts.*

By *Ven. Mandey*, Philomat.

LONDON,

Printed for *Tho. Shelmerdine* at the *Rose Tree*, and *Tho. Ballard*, at the *Rising Sun* in *Little-Brittain*, 1702. Where is to be Sold, the *Universal Mathematical Synopsis*, by the same Author. Price 6 *s*.

NO. 19

The first book in English to treat of the technicalities of paper maceration. This title-page is from the second edition of 1702, but the text of the book is identical with that of the first edition of 1696 until 1847. The verse relates to William Bradford and the Rittenhouse paper mill:

> Here dwelt a printer and I find
> That he can both print books and bind;
> He wants not paper, ink, nor skill
> He's owner of a paper mill.
> The paper mill is here hard by
> And makes good paper frequently,
> But the printer, as I do here tell,
> Is gone into New York to dwell.
> No doubt but he will lay up bags
> If he can get good store of rags.
> Kind friend, when thy old shift is rent
> Let it to th' paper mill be sent.

Another work dealing slightly with Pennsylvania papermaking appeared two years after John Holme's verse, or in 1698, with the title, *An Historical and Geographical Account of the Province and Country of Pennsylvania.* This was published in London, and the reference to paper is: "All sorts of very good paper are made in German-Town, as also very fine German linen such as no person of quality need be asham'd to wear."

One of the rarest of papermaking items is a small octavo volume which was printed in Hamburg in 1699. The title-page reads, *Die Preiss-würdige Pappiermacher-Kunst, Oder: Gründliche Beschreibung, Was für eine edle Kunst und theure Gave GOTTES es sey umb die höchst-nöthige Pappiermacher-Kunst, wie es aus GOttes Wort und denen Historien, wie auch andern dienlichen Gründen, mit der täglichen Praxi und Erfahrung, ec. zu beweisen.* This little book has but 96 pages and no illustrations.

In 1699 a letter of papermaking interest was written by Humphrey Wanley (1671–1726), to Samuel Pepys, the diarist. Wanley was custodian of heraldic manuscripts to Robert Harley, Earl of Oxford. The letter is in regard to a collection of paper that Wanley

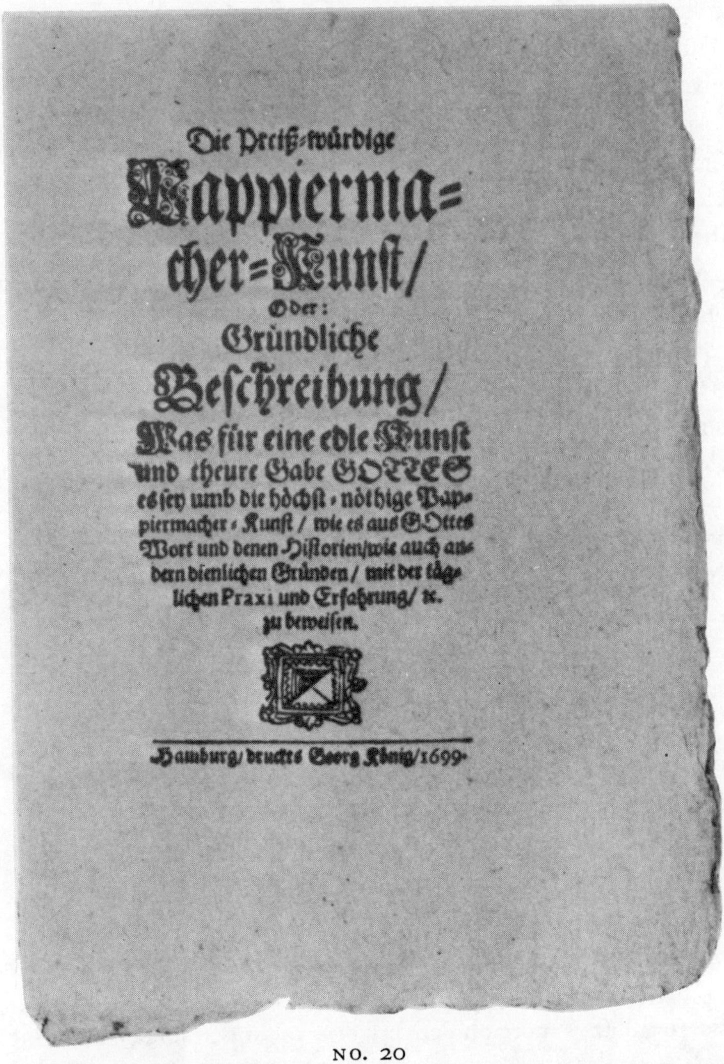

Die Preiß-würdige
Pappierma-
cher-Kunst/
Oder:
Gründliche
Beschreibung/
Was für eine edle Kunst
und theure Gabe GOTTES
es sey umb die höchst-nöthige Pap-
piermacher-Kunst / wie es aus GOttes
Wort und denen Historien/wie auch an-
dern dienlichen Gründen / mit der täg-
lichen Praxi und Erfahrung/ rc.
zu beweisen.

Hamburg/ druckts Georg König/ 1699.

NO. 20

Title-page of the rare late seventeenth century German book on the art of papermaking

had formed, and reads: "This collection of shreads and scraps of old parchment makes me bold to mention to you, Sir, another like collection of paper that I have made when I was last in London. This I have put together in two volumes of which I venture to give you some account. (The two volumes, bound in one, now form the Sloane Manuscript No. 526, in the British Museum.) The first volume contains: Some of our countrey common brown, grey, blue, red, and black paper. Some Eastern paper written upon in the East Indies, Persia, Turkey, and China, and some Chinese paper printed upon . . . The other volume contains a collection of sheets and pieces of white European paper which are plac'd according to their several marks and stamps. The reason why I gathered them together was not only to satisfy my curiosity, in knowing what sort of paper is in use among so many nations, or to please myself or my friends with looking on fine colours, etc., but to know whether they would confirm a notion I have of the originall of our paper, which, to me, they do. And further, I thought that some time or other the history of paper might be attempted, and with so much the more hope of success, when the author, whoever he may be, has before his eyes most of the sorts of that subject that he is to treat about."

The writer of this letter, Humphrey Wanley, had opportunity further to aid in the collection of papers, as will be noted in the next item, relating to John Bagford.

There has never been a more ardent collector than this John Bagford (1675–1715), even though his pursuits were not always legitimately carried on. In 1707 Bagford published in the Philosophical Translations: "A proposal for a history of printing, printers, illuminators, and papermaking; on subscription 10 shillings, and 10 shillings more on delivery of a volume in folio, containing 200 sheets." The book was never published, but the manuscripts and material that Bagford possessed are now in the British Museum, bearing the numbers 5891–5988. After Bagford's death the manuscripts and material were purchased by Humphrey Wanley of Lord Oxford's Library and in due course went to the British Museum with the Harleian Manuscripts, the volumes now being

NO. 21
John Bagford, "the wicked old biblioclast"

known simply as "Bagford's Collectanea." In volume 5940 there are specimens of end-papers and a number of watermarks from the sixteenth century. In volume 5941 there may be seen samples of Oriental papers, while in volume 5942 there are examples of paper with imprints of the makers and printed advertisements relating to the manufacturing of paper. There is not a great deal of paper interest in the Bagford collection, but he has always been considered the great authority of his time on paper, just how justified this was it is not now possible to say. William Blades in *The Enemies of Books* calls Bagford "a wicked old biblioclast," as he was in the habit of visiting libraries and removing illustrations, end-papers, or whatever struck his fancy from the public books, so that he might add the items to his collection! Bagford died in May, 1715.

The first work to mention papermaking materials, other than linen and cotton, appeared in 1716 in a small octavo volume under

NO. 22

The engraving used by "The Society of Gentlemen" to show the preparation of hemp for a papermaking material. This book, 1716, was the first to suggest a substitute for linen and cotton

the title of: *Essays, for the Month of December* 1716, *to be continued Monthly, by a Society of Gentlemen. For the benefit of the People of England.* The work was printed in London and the part regarding papermaking may be found in Essay VI. "The Society of Gentlemen" advanced the idea of using raw hemp without spinning or weaving, as a material from which to fabricate paper, and they go into detail as to its preparation for the purpose. They suggest also that paper mills be set up on barges or flat-bottomed boats on the Thames and the other navigable rivers of England, as they are on the Elbe and the Danube, the power to be furnished by the flux and reflux of the tide. There is an engraving which shows how the hemp may be washed, boiled, pressed, and dried so

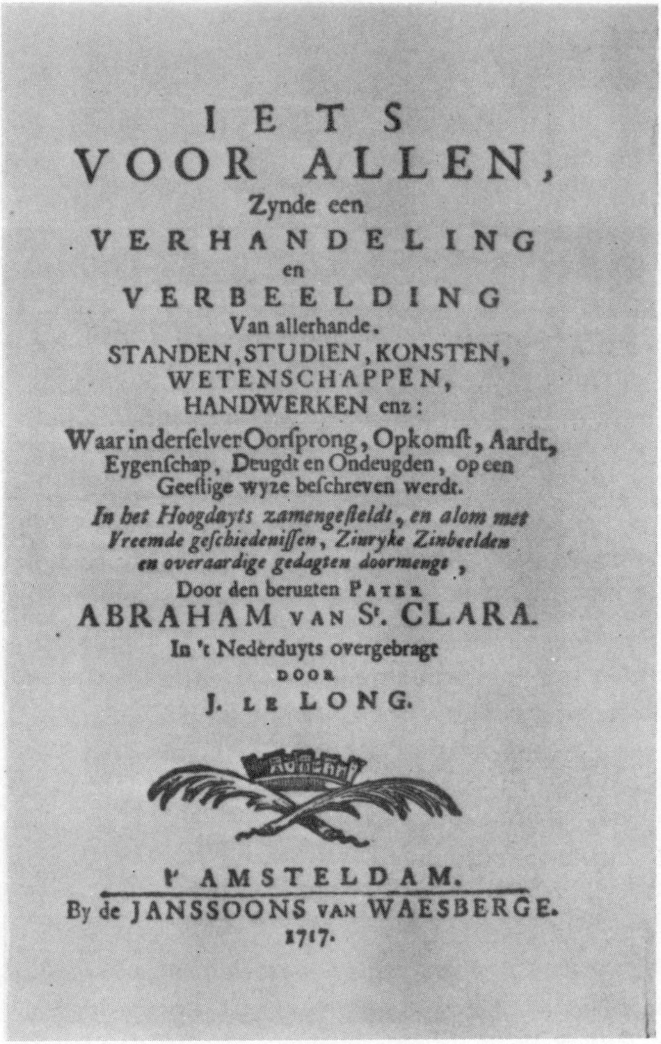

IETS
VOOR ALLEN,
Zynde een
VERHANDELING
en
VERBEELDING
Van allerhande.
STANDEN, STUDIEN, KONSTEN,
WETENSCHAPPEN,
HANDWERKEN enz:
Waar in derſelver Oorſprong, Opkomſt, Aardt,
Eygenſchap, Deugdt en Ondeugden, op een
Geeſtige wyze beſchreven werdt.
In het Hoogduyts zamengeſteldt, en alom met
Vreemde geſchiedeniſſen, Zinryke Zinbeelden
en overaardige gedagten doormengt,
Door den berugten Pater
ABRAHAM van St. CLARA.
In 't Nederduyts overgebragt
door
J. le LONG.

t' AMSTELDAM.
By de JANSSOONS van WAESBERGE.
1717.

NO. 23

Title-page of the book of trades compiled by Abraham van St. Clara, the writer of the well-known lines under the title of, "Was ist ein Papier?"

NO. 24
Abraham van St. Clara

that it can be put to use as a substitute for linen and cotton; it is suggested that the paper mill owners plant hemp in their mill yards and gardens and produce their own supply of papermaking material.

One of the prettiest little engravings showing a papermaker at work is in the Dutch book of trades compiled by Abraham van St. Clara (1644–1709) and published in Amsterdam in 1717, with a title-page in red and black, reading: *Iets Voor Allen, Zynde een*

NO. 25

The papermaker, from the book of trades by Christoff Weigle, 1698

Verhandeling en Verbeelding van Allerhande, Standen, Studien, Konsten, Wetenschappen, Handwerken, etc. This book of trades devotes five pages to the art of papermaking, and presents a well-executed engraving of the vatman forming a sheet of paper. Abraham van St. Clara was a Roman Catholic priest of Vienna and his earliest writing regarding paper appeared about the year 1700 under the caption, *Was ist ein Papier?* The manuscript reads: "Ein Papier ist ein Werkzeug der Gelehrten, ist ein Vorrat der Bücher. Ein Papier ist ein Hausrat der Kanzlei, ist ein Schatz der Schüler, eine Erhalterin der menschlichen Freundschaft; ein Papier ist der grösste Nutz der Welt. Mein Papier, wer sind, bekenne es recht, deine Eltern gewest? Mein Vater, sagtes, war ein Lumpen, meine Mutter ein Fetzen. Von einem so schlechten Lumpengesind eine so herrliche Sach!" In 1718, there was published at Amsterdam another little book of trades, entitled: *Spiegel van het Menselyk Bedryf, vertoonende honderd verscheiden Ambachten, Konstig afgebeeld, en met Godlyke Spreuken en Stitchtlyke Verzen verrykt.* This book had been compiled by Jan En Kasper Luiken (1670–1710), and one of the plates shows a papermaker. The volume was founded on the earlier work of Abraham van St. Clara and the Luiken plate depicting papermaking is merely the plate of the earlier work reversed. The engraving was not originally made for Abraham van St. Clara's book but was probably used for the first time in the book by Christoff Weigle in 1698, with this verse beneath:

> Der alte Lumpe kommt durch Fleiss
> zu neuen Nutzen schön und weiss:
> Solst du mein Hertz, verächtlich bleiben?
> Hervor aus altem Sünden Stand,
> ganz neu und rein dass Gottes Hand
> auff dich mög seinen Willen schreiben.

The next book in chronological order would be the important work of Leonhardt Christoph Sturm (1669–1719), but inasmuch as this volume is taken into consideration in Part III on the subject of papermaking maceration, it will not be necessary to describe the work here.

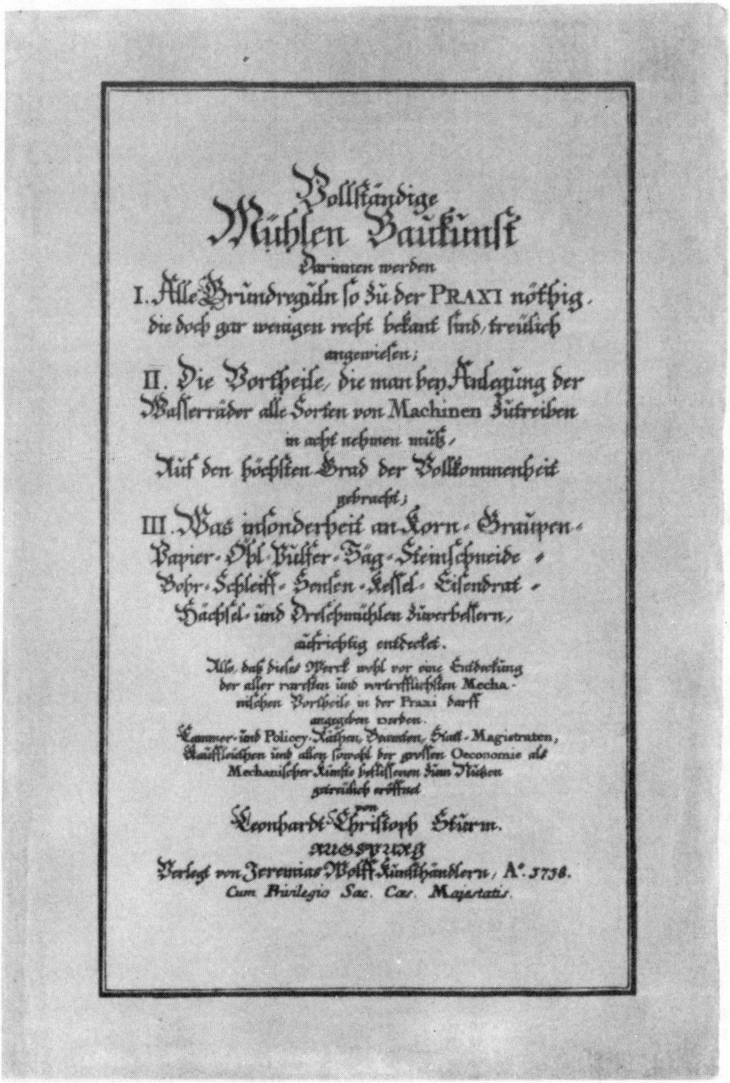
Vollständige
Mühlen Baukunst
Darinnen werden
I. Alle Grundregeln so zu der PRAXI nöthig, die doch gar wenigen recht bekant sind, treulich angewiesen;
II. Die Vortheile, die man bey Anlegung der Wasserräder alle Sorten von Machinen zutreiben in acht nehmen muß, Auf den höchsten Grad der Vollkommenheit gebracht;
III. Was insonderheit an Korn- Graupen- Papier- Öhl- Pulver- Säg- Steinschneide- Bohr- Schleiff- Sensen- Kessel- Eisendrat- Häcksel- und Dreschmühlen zuverbessern, aufrichtig entdecket.
Also daß dieses Werck wohl vor eine Entdeckung der aller raresten und vortrefflichsten Mechanischen Vortheile in der Praxi darff angegeben werden.
Cammer- und Policey-Räthen, Beamten, Statt-Magistraten, Kauffleuthen und allen sowohl der grossen Oeconomie als Mechanischer Künste beflissenen zum Nutzen getreulich eröffnet
von
Leonhardt Christoph Sturm.
AUGSPURG
Verlegt von Jeremias Wolff Kunsthändlern, A°. 1718.
Cum Privilegio Sac. Caes. Majestatis.

NO. 26

The title-page of the book on mill machinery by Leonhardt Christoph Sturm

The use of wood as a material from which to make paper was suggested by René Antoine Ferchault de Réaumur (1683–1757), a celebrated naturalist and physicist residing in France. Réaumur had observed the habits of the wasp and concluded that the wood material which was used by these insects to construct their nests that so much resembled paper, could also be used in the actual process of papermaking. Réaumur's observations regarding the habits of the wasp were laid before the French Royal Academy in a treatise dated November 15, 1719. Réaumur wrote: "The American wasps form very fine paper, like ours; they extract the fibres of common wood of the countries where they live. They teach us that paper can be made from the fibres of plants without the use of rags and linen, and seem to invite us to try whether we cannot make fine and good paper from the use of certain woods. If we had woods similar to those used by the American wasps for their paper, we could make the whitest paper, for this material is very white. By a further beating and breaking of the fibres which the wasps make and using the thin paste which comes from them, a very fine paper may be composed. This study should not be neglected, for it is, I daresay, important. The rags from which we make our paper are not an economical material and every papermaker knows that this substance is becoming rare. While the consumption of paper increases every day, the production of linen remains about the same. In addition to this the foreign mills draw upon us for material. The wasp seems to teach us a means of overcoming these difficulties." Réaumur further continued his observations regarding a heavier wasp-made paper: "But all of the wasps of the Kingdom that I know make nothing as singular as a species of wasp that lives in Canada . . . At first glance, and even after examining the surface for a considerable time, one would accept the nest as the work of the hand of man. Its covering resembles our paper to such an extent that it is hard to detect a difference. It has the same gloss and the colour is that of an old piece of manufactured paper which had formerly been white. It is fine paper and as heavy as that of ordinary portfolios."

Réaumur's most important observations relating to natural history may be found in *Mémoires pour servir à l'histoire des in-*

NO. 27

René Antoine Ferchault de Réaumur, the first scientist to suggest the use of wood as a papermaking material

sectes," published in Amsterdam, six volumes, 1737–1748. There were to have been twelve volumes of this work, but the author died during the process of compiling the seventh volume. There is no record of Réaumur's ever having actually made paper from wood, but he was, no doubt, the first scientist to suggest or recommend its use for this purpose. Human invention had, in the case of paper, been anticipated by the wasp, which insect may be considered as a professional papermaker, devoting most of her time and energies to the fabrication of this material which she uses in the construction of nests. For this purpose the wasp seeks dry wood,—fence-rails and weather-beaten boards are a favourite source of supply. This substance she saws or rasps by mastication, mixes the macerated material with a natural size which is exuded for the purpose, and, working the whole into a paste, spreads the paper substance in a manner truly remarkable. The nest is usually a prolonged irregular spheroid, exceptionally light in weight, of a dark grey colour, and bound with repeated bands of paper to the bough from which it is suspended. The nest is water resistant to a high degree, partly from the rounded top, but more from the fact that the paper strips of which it is composed overlap, like the shingles of a house, and from the fact that a semi-waterproof size is used in the composition. The community of papermaking wasps may amount to thirty thousand in a single season, according to Réaumur. It is interesting to note Réaumur's condemnation of himself for having neglected to attempt any experiments on his own account in the making of paper from wood. In 1742 his comment was: "I am ashamed that I have not yet tried this experiment, it is more than twenty years since I first realized its importance and made an announcement of it. But I had hoped that some one would have been interested in making it his occupation."

The investigations of Réaumur, while not in actual papermaking, gave the hint to European scientists that paper might be made from other substances than rags. In 1727–1730, Franz Ernst Brückmann (1697–1753), compiled a book on geology,[1] a few copies of

1 Magnalia Dei in locis svbterraneis, oder Unterirdische schatzcammer aller königreiche und länder, in ausführlicher beschreibung aller, mehr als MDC. bergwercke durch alle vier welt-theile . . . nebst anmerckung aller derjenigen länder und oerter, wo edelgesteine zu finden, in geographischer ordnung und einigen kupfferfiguren zu besichtigen dargestellet von Francisco Ernesto Brvckmann. Braunschweig, 1727–30. 2 volumes. 22 x 35 cm.

which were printed on paper made of asbestos, the first use of asbestos for this purpose. Albert Seba (1665–1736), a Flemish writer on natural history, compiled a set of books[1] in 1734–1765, in which he has the following to say regarding material from which paper

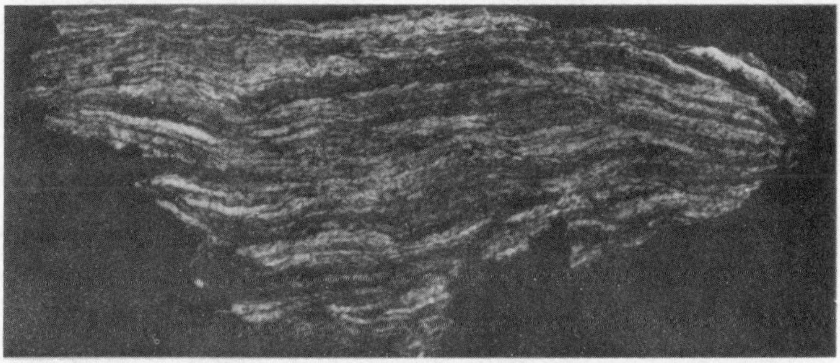

NO. 28
The wasp was the first papermaker

could possibly be made: "My country does not seem to lack trees suitable for making paper, if people would give themselves the necessary expense and trouble. 'Alga Marina' for example, which is composed of long, strong viscous filaments, might it not be proper for this purpose, as well as the mats of Muscovy, if they were prepared as the Japanese make their timber?"

In 1741, Jean Étienne Guettard, a French physician (1715–1786), made his first observations regarding substitutes for rags in papermaking and wrote several articles[2] in French advocating the use

1 Locupletissimi rerum naturalium thesauri accurata descriptio, et iconibus artificiosissimis expressio, per universam physices historiam. Opus, cui, in hoc rerum genere, nullum par exstitit. Ex toto terrarum orbe collegit, digessit, descripsit, et depingendum curavit Albertus Seba. Amstelaedami, 1734–65. 4 volumes. 35 x 53 cm.

2 Observations sur différentes matières dont on fabrique le papier. In: Mémoires de Paris. Paris, 1741.
Mémoire sur les différentes manières de fabriquer le papier. In: Histoire de l'Académie des sciences, année 1741; Journal économique. July and August, 1751.
Recherches sur les matières qui peuvent servir a faire du papier. In: Mémoires sur différentes parties des Sciences et Arts. Volume 1. Paris, 1768.

NO. 29

Albert Seba, a Flemish writer on natural history, who recommended the use of "Alga Marina," a sort of seaweed, as a material from which to make paper

SELECT ESSAYS:

CONTAINING:

The Manner of raising and dressing Flax, and Hemp.

Also, The whole Method of Bleaching or Whitening Linen-Cloth.

Likewise, Observations on the Management of Cows and Sheep.

The Manner of raising Radishes, Turnips, Cabbage, and other such Plants.

And an Enquiry, concerning the materials that may be used in making Paper.

With, Valuable Dissertations on other useful Subjects.

Collected from the Dictionary of Arts and Sciences, and from various modern Authors.

Felix, qui potuit rerum cognoscere causas.

Happy, he who can trace out the Causes of Things.

Georgic 2. Davidson's Virgil.

PHILADELPHIA:

Printed, by ROBERT BELL, next Door St. Paul's Church, in Third-street, Philadelphia.

M,DCC,LXXVII.

NO. 30

Title-page of the earliest American book dealing with materials for papermaking. Philadelphia, 1777

of Conferva as a papermaking material. One of the articles is accompanied by actual specimens of papers made from the bark, leaves, and woods of various trees, plants, and shrubs. Guettard's observations were translated into English in 1754 and published in London.[1] On pages 372–406 of this volume on miscellaneous subjects Guettard recommends the use of various forms of vegetation for papermaking. He had not, however, at this time carried his ideas to any conclusion, for on page 406 of his essay he says: "Such examinations have always constituted my desires, since I thought of making experiments upon paper. I have been unable, hitherto, to accomplish my scheme; but, nevertheless, I have reason to hope I shall one day see it accomplished." In 1777 there was published in Philadelphia a small octavo volume[2] which contained the treatise by Guettard that had been published in London in 1754. This book is unusually interesting as it is the first publication issued in America which treated of papermaking materials.

John Strange (1732–1799), an English diplomatist and traveller wrote a pamphlet[3] in 1764 which was issued in Italy, relating to papermaking materials. This treatise was compiled while the author was travelling in Italy and he tells about the paper of Cortona, a town of Tuscany, and gives a summary of the use of broom, a fabaceous shrub, and other plants for papermaking. While Strange recommends Conferva for various uses he does not deem it suitable material from which to make paper. The researches made by John Strange are vague and unimportant.

1 Inquiry concerning the materials that may be used in making paper. By Mr. Guettard, of the Royal Academy of Sciences, and Physician to his Serene Highness the Duke of Orleans. In: Select Essays on Commerce, Agriculture, Mines, Fisheries, and other useful subjects. London, 1754. 14 x 23 cm.

2 An enquiry concerning the materials that may be used in making paper. In: Select Essays: Collected from the Dictionary of Arts and Sciences, and from various modern Authors. Philadelphia, 1777. 11 x 19 cm.

3 Letter on the origin of the natural paper of Cortona, with other observations relative to the uses and excellent qualities of the Conferva of Pliny. Pisa, 1764.

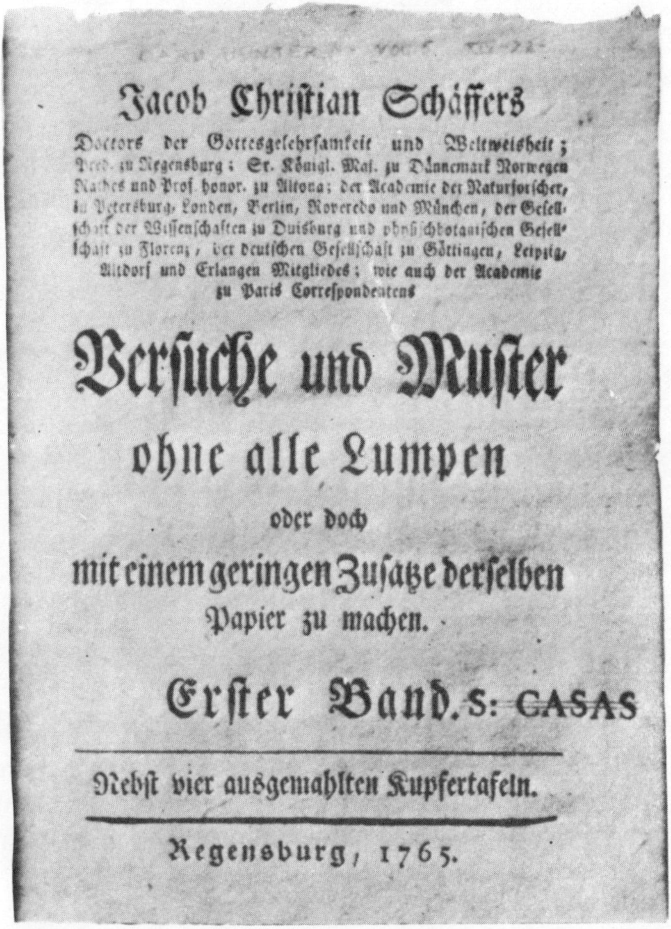

Jacob Christian Schäffers

Doctors der Gottesgelehrsamkeit und Weltweisheit; Pred. zu Regensburg; Sr. Königl. Maj. zu Dännemark Norwegen Rathes und Prof. honor. zu Altona; der Academie der Naturforscher, zu Petersburg, London, Berlin, Roveredo und München, der Gesellschaft der Wissenschaften zu Duisburg und physischbotanischen Gesellschaft zu Florenz, der deutschen Gesellschaft zu Göttingen, Leipzig, Altdorf und Erlangen Mitgliedes; wie auch der Academie zu Paris Correspondentens

Versuche und Muster

ohne alle Lumpen

oder doch

mit einem geringen Zusatze derselben

Papier zu machen.

Erster Band. S: CASAS

Nebst vier ausgemahlten Kupfertafeln.

Regensburg, 1765.

NO. 31

Title-page of the first volume of Dr. Schäffer's six volume treatise on papermaking materials

We now come to the highly interesting and unusual work of Jacob Christian Schäffer (1718–1790), who did more than any of his predecessors in the quest for materials for papermaking. In

1765 he commenced a treatise[1] on the subject which was issued in six volumes, the last appearing in 1771. This is the rarest work on the specific subject of paper that has ever been published, but five or six copies being known in America. Schäffer was born in Querfurt, near Merseburg in Saxony in the year 1718, but his family early removed to Regensburg, or Ratisbon, the capital of the Bavarian province of Oberpfalz, about one hundred and fifty miles south of his birthplace. Schäffer studied for the ministry and became a well-known clergyman in Regensburg, but it seems that he derived his greatest pleasure outside the Church, as he devoted most of his time to the study of science and natural history. His work on the fungi of Bavaria is still regarded as a standard authority. Schäffer's interest in the flora of Bavaria directed his attention to the possibility of new materials for papermaking, and it is with his researches in this direction that we are concerned. In his six volume treatise Dr. Schäffer has left a permanent record of his experiments in the search for new papermaking materials, and the actual specimens of his paper establish the fact that he was the pioneer in the use of many vegetable fibres for the fabrication of paper. It was not Schäffer's desire, as he explains, to make well-finished paper; he simply wished to show the vast variety of vegetation available for the purpose. As his experiments were carried on previous to the discovery of bleach, his examples of paper have the tint of the original materials from which they were made. In most of the samples about one-fifth part cotton rags were added to the pulp to help bind the fibres together. A number of the specimens are sized, and nearly all have been printed upon, showing from what plant or fibre they had been made.

It is curious to note that one of the first specimens shown in Schäffer's books was made from wasps' nests,—for was not the

1 Versuche und Muster ohne alle Lumpen oder doch mit einem geringen Zusatze derselben Papier zu machen. Regensburg, 1765–71. 6 volumes. 13 x 18 cm.

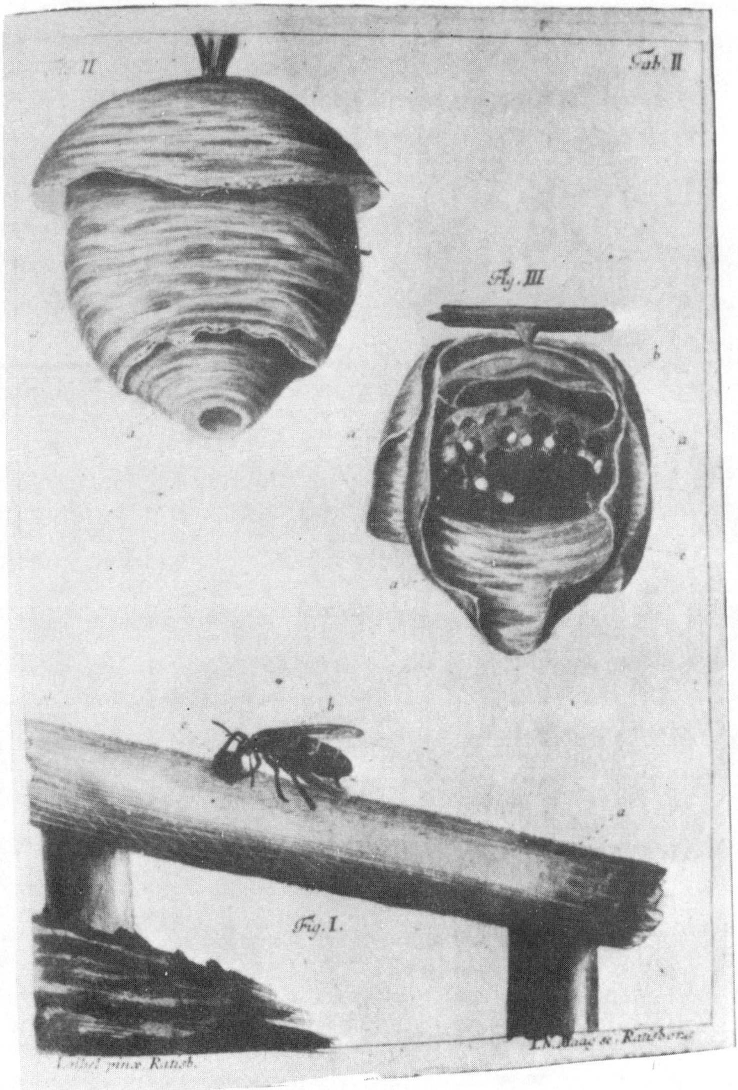

NO. 32

It was by studying the work of the wasp that Réaumur and Schäffer conceived the idea of making paper from wood. From Dr. Schäffer's treatise

wasp, as Réaumur pointed out, the first papermaker? Dr. Schäffer's researches extended over a period of more than eight years, and all of the materials with which he worked were gathered in his own garden or neighbouring fields. The mode of making the vari-

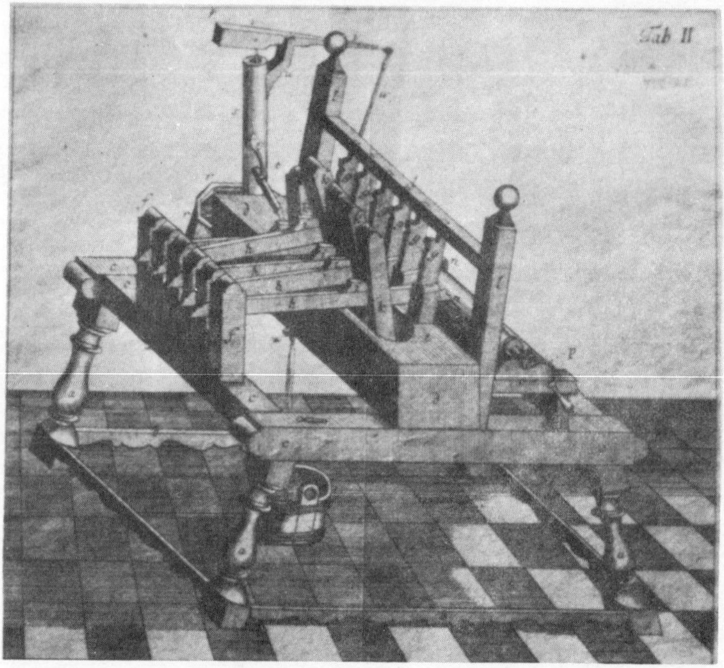

NO. 33

The small, hand-operated stampers that were used by Dr. Schäffer in making his experiments with various fibres

ous samples of paper is given in detail. The materials, according to their nature, were first chopped by hand, the different kinds of wood having been reduced by a toothed plane. Most of the experiments were carried on in Dr. Schäffer's own home, and for the purpose of macerating the materials he employed a miniature set of stampers operated by hand. In most cases the vegetable matter was subjected to the stampers without preliminary treatment, but with a few of the materials milk of lime was used. With some of the more stubborn fibres the Doctor used a stiff lime paste and suffered the material to remain in it for various lengths of time

NO. 34
Dr. Jacob Christian Schäffer

and he noted that this treatment reduced the beating process and gave a brown tint to the wood.

All of the samples in the Schäffer books were formed on a "laid" mould which measured about eleven by fifteen inches, and an indistinct watermark of a crest appears on most of the specimens. The paper used for the text of the volumes bears a watermark of crossed-keys with the initials C M M inside a border. There are not many copies of this work extant and the specimens in the various sets are not always identical. The compilation is more often found bound complete in one volume, than in the six original pamphlets as Dr. Schäffer issued his work. The six volumes are designated on the title-pages as volumes one, two, one, two, three, and four. The first volume of this valuable work was completed

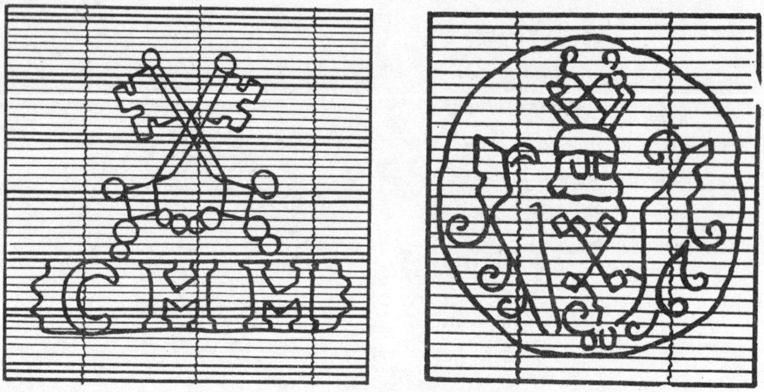

NO. 35

The C M M watermark appears in the text pages of Dr. Schäffer's six volumes, while the crest was used as a papermark in the mould upon which nearly all of the specimens were fabricated

January 30, 1765, and the book contains fifty-five pages of text, five plates and fifteen specimens of paper. The examples include papers that were made from wasps' nests, and from various kinds of wood, moss, and vines. In volume two, dated April 3, 1765, there are twenty-eight pages of text and one plate. The specimens embrace papers that were fabricated from hemp, bark, straw, and cabbage stalks. The third volume (in the order of their appearance), is dated November 3, 1765, has thirty-two text pages, and the specimens include papers made from asbestos, cat-tail and burdock stalks, thistles, and turf. Volume four was completed January 1, 1766, with twenty-four pages of text and eleven samples, including papers from seed, mallow, St. John's wort, and Indian corn husks. The fifth volume is dated April 15, 1767, while the sixth, and last, is dated 1771. These two books give specimens of paper that was

NO. 36
The specimen from Dr. Schäffer's book made from roof shingles and painted upon to show the adaptability of the paper for this purpose

made from genista, pine-cones, potatoes, old shingles, reeds; and bean, horse-chestnut, walnut, tulip, and linden leaves; also paper made from yellow and brazil wood.

Inasmuch as Dr. Schäffer's work is one of the most important works that deal with paper, and owing to the text being remarkably interesting, it will not be out of the way to quote at length from the book. I am indebted to Dr. Zoltan Haraszti for the translation which follows: "It is generally known," writes Schäffer, the clergyman, "that the paper which, according to all evidence, has been used in Europe since the twelfth century, is made of rags and worn-out linen. And the dearth of this material is now complained of everywhere. The most curious thing is that not only a certain kind of paper is wanting; statements of merchants reveal that in regard to wrapping paper, card-board, etc., the want is even keener.

"This general lack of paper, and the harm done thereby to administration, science, and commerce, brought a few years ago to my memory what various scholars—like Seba, Réaumur, Guettard, Gleditch, and others—had in mind and proposed in regard to papermaking. They believed and with probability proved, that one is not exclusively bound in papermaking to linen, but can make paper just as well of a great many other things. It is generally known that rags originally were made of lint and hemp, which are plants themselves; thus, these scholars came to the conclusion that every material which—like hemp and lint—consists of such soft, elastic, easily-separated fibres as through the action of water turn to pulp and, by drying, attain a certain stiffness and firmness, must be fit for papermaking.

"Few objections can be reasonably raised against the statements of these scholars,—and the more certain it is that besides hemp and lint there are many plants having the same necessary qualities, the more difficult it is to comprehend why these ideas have not been used for the public benefit, and why such experiments have not been longer, oftener, and in a more satisfactory way pursued. This regrettable neglect induced me three years ago to get to work

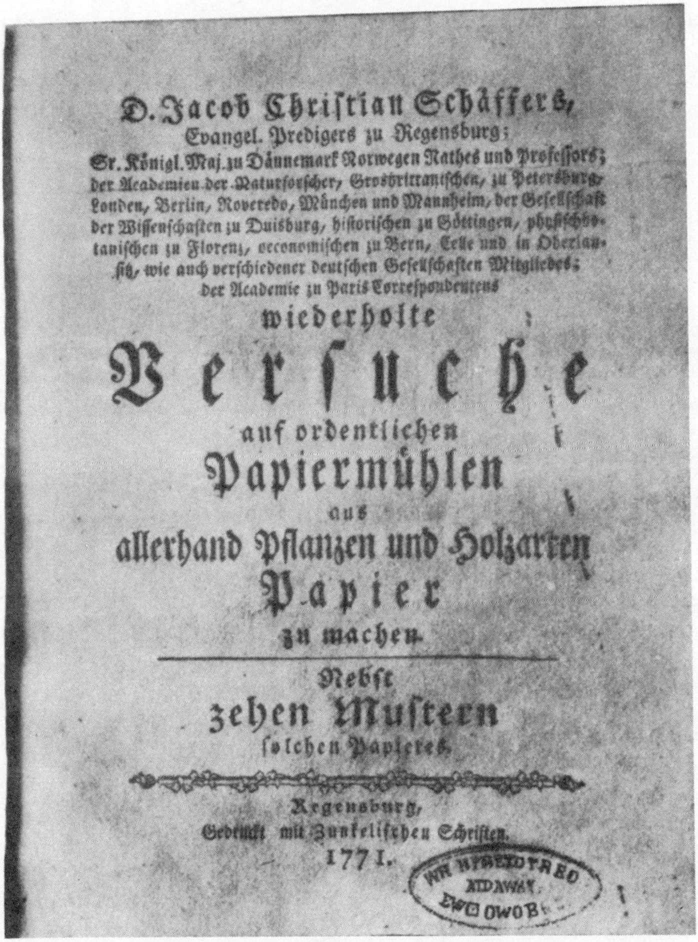

D. Jacob Christian Schäffers,
Evangel. Predigers zu Regensburg;
Sr. Königl. Maj. zu Dännemark Norwegen Rathes und Professors;
der Academien der Naturforscher, Grosbritannischen, zu Petersburg,
London, Berlin, Roveredo, München und Mannheim, der Gesellschaft
der Wissenschaften zu Duisburg, historischen zu Göttingen, physischbo-
tanischen zu Florenz, oeconomischen zu Bern, Celle und in Oberlau-
sitz, wie auch verschiedener deutschen Gesellschaften Mitgliedes;
der Academie zu Paris Correspondentens
wiederholte
Versuche
auf ordentlichen
Papiermühlen
aus
allerhand Pflanzen und Holzarten
Papier
zu machen.
Nebst
zehen Mustern
solchen Papieres.
Regensburg,
Gedruckt mit Zunkelischen Schriften.
1771.

NO. 37
Title-page of the sixth volume of Dr. Schäffer's work. This is the rarest book of the set

with all possible energy. It seemed to me as if nature itself wished to encourage me in my task. Taking a walk outside of our city, chance led me to a place where one side of the field, from abundance of pappus of the poplar, and the other from "wool-blade," looked wholly white. At this sight the thought flashed through my mind: Could not paper be made from these plants?

"Without losing time, I started at once to experiment. I gathered the pappus as well as the wool-blade, then talked the matter over with the papermaker of the town, Herr Mechenhäuser. How happy I was when—after examining the pappus,—this good-natured man declared that, though the wool-blade did not seem to him fit for papermaking, the poplar-downs must by all means be tried out. But my joy was gone when I learned that the papermaker wanted five to twenty pounds of pappus. It was impossible to get such a quantity, and it cost me a great deal of talk before I could persuade him to make the experiment with the poplar-downs in a mortar. A few days later I received a few samples of the new paper. It was, at any rate, paper; one could print or even write on it. Only, it was too ragged, and did not possess the necessary firmness, and, besides, it was full of little brown knots, the residue of the pounded kernels. Yet these first and imperfect samples provided the most convincing proof that the pappus of poplar is fit for papermaking; and the papermaker assured me, that if a satisfactory quantity of pappus had been pounded in his regular beater instead of in the mortar, and if it had been further duly treated and finished, we could have obtained a reliable and usable paper.

"Other occupations prevented me continuing my experiments in the following two years. Yet being urged from several sources, especially the Academy, I resumed the experiments. I gathered again a basketful of poplar-pappus and wool-blade, and gave it to the same papermaker. After a while I obtained paper from both species. The wool-blade paper came out unusually poor. It was extremely brittle,—all expense and labour seemed to be spent in vain. Nevertheless, I wondered whether there could not be made better paper of wool-blade. It was evident that the wool lay too

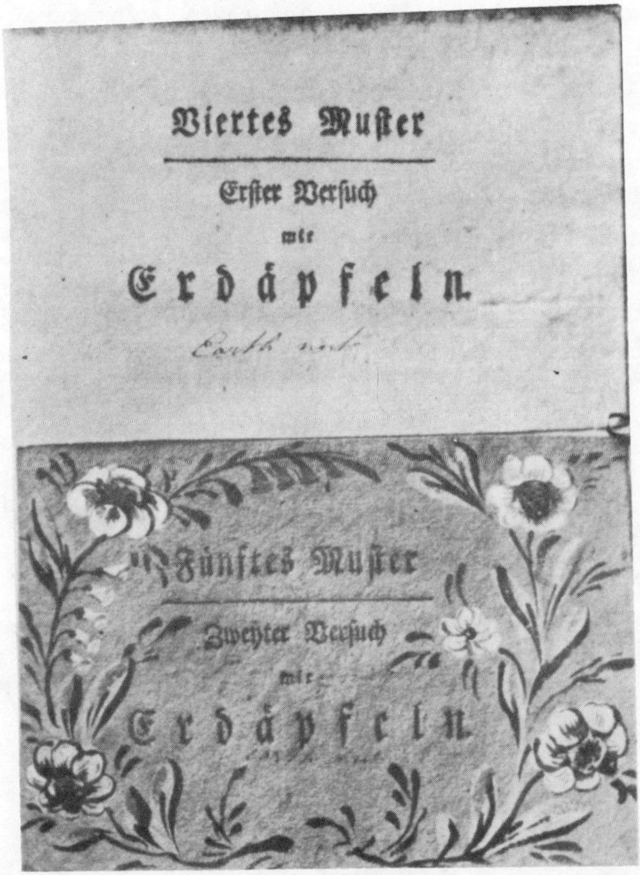

NO. 38

Two specimens in the fifth volume of the work by Jacob Christian Schäffer. The upper example was made from the insides of potatoes, the lower from the skins

long in the lime, and its excessive brittleness and rigidity were caused by the lime and not by the nature of the material. I was anxious to know whether the next experiment supported my supposition. The paper made of pappus of poplar was incomparably better. It was a perfectly good writing paper, lacking nothing but a little more white to be accepted by everybody as rag paper. I noticed too, that the sheets differed in colour. I now selected three different kinds. The first was dirty-white, the second grey, and the third yellowish. Could this difference be caused by the pappus itself? This seemed unlikely to me. I thought rather that the cause must be sought in the process of making. And I was not mistaken. For, to my questions, the papermaker avowed that he worked at three different times on the paper, and consequently one portion lay longer in the lime than the other. Could I not conclude from this that the grey and yellow colours had accidentally developed and that with better care a better quality could be produced?

"I honoured myself by submitting three new and good samples to the Bavarian Academy, and without delay, I made preparations for experiments with all of the materials which I thought might be fit for papermaking. And since my former experience convinced me that my aim would be very slowly and with double expense—if at all—reached if I depended on the papermaker alone, I decided myself to make all the experiments, from beginning to end, in my own home."

After Doctor Schäffer decided to be his own papermaker he employed a local artisan to show him the principles of the craft, also his servants were instructed, and after procuring several moulds he was ready to make his own experiments. The clergyman-papermaker had a small stamping mill constructed which was operated by hand, and in this machine, which he pictures in his book, most of the beating was accomplished for producing the samples of papers shown in the six volumes. "And I started again to make paper," concludes the story of the ingenious Schäffer, "from poplar-downs, and shavings, and sawdust, and so forth. And what sweet satisfaction did I feel when I saw that everything came out better

NO. 39

The frontispiece of the six volume work by Dr. Jacob Christian Schäffer showing the various processes of papermaking

than I had imagined! In a short time I was able to produce a reliable new sort of paper. After such success, how could I forbear to make new and again new experiments? —especially when the cost of the tools and the studying of the process had already been met. I decided, therefore, to make these experiments my regular winter occupation. I could do so the more readily since the paper made hitherto found the warmest acknowledgment and I was urged to continue my experiments."

In making the specimens for his books Doctor Schäffer frequently found that the same plant, collected at different times, gave different results, which is evident in an examination of the specimens. Experiments were made in the use of potatoes for papermaking and it was found that sheets of paper could be fabricated from both the skins and the insides of this tuber. However, when the good Doctor was carrying out his researches in the use of potatoes he could not that year (1767) procure enough of these vegetables for his work. In fact, potatoes were so little known in his country at that time that Doctor Schäffer felt that he need explain their use and wrote the following regarding them: "The earth-apple (potato) is a kind of plant known principally in Voigtland, France, Austria, and since some years also in Bavaria and the Pfalz, on whose fibrous roots there are produced, in the earth, uneven various-sized knotty and apple-like growths, which are known as earth-apples. These earth-apples are an uncommonly useful vegetable for the kitchen and for all housekeeping, particularly where there is a scarcity of bread. Poor people in many places not only really eat these earth-apples, but the appetite is satisfied by them as well as by bread."

The three sets of Schäffer's works in my collection are bound, each volume separately, in the original full calf with gold-tooled backs. In a complete set there should be 176 pages of text with ninety-five specimens of paper, two examples of paper lace, four

Dr. JACOB CHRISTIAAN SCHÄFFERS

PROEFNEMINGEN

EN

MONSTER-BLADEN,

OM

PAPIER TE MAAKEN

ZONDER LOMPEN,

OF

Met een gering Byvoegzel derzelven.

TWEEDE DEEL.

Uit het Hoogduitsch vertaald.

Met Achttien Papier-Monsters en Eene gekleurde Plaat.

Te AMSTERDAM,

By JAN CHRISTIAAN SEPP, Boekverkoper.

MDCCLXX.

NO. 40

Title-page of the rare Dutch edition of Dr. Jacob Christian Schäffer's book on the use of plants and barks for making paper

samples of paper cloth, and fifteen engravings, a few of which have been coloured by hand. A few years ago an incomplete copy of Schäffer's work sold at auction for four hundred and twenty-five dollars, but there were only four parts to this set and they had been rebound in two volumes, also this set lacked thirty-four specimens of paper.

The researches of Jacob Christian Schäffer were evidently considered of importance at the time, for in 1770, or even before Schäffer's work was completed, there appeared in Amsterdam a book regarding his experiments. This book,[1] almost rarer than the Regensburg edition, comprises 32 pages of text, and one hand-coloured engraving of various forms of vegetation, including cat-tails, earthmoss, and "wilde wyngaard." There are eighteen specimens of paper, which embrace sheets made from nettles, straw, earthmoss, cat tails, and aloe leaves.

In the eighteenth century, as may be seen from the foregoing works, there was considerable research and interest in finding new material, other than rags, from which to make paper. While there were numerous specimens of papers which had been made from newly-discovered materials, the books of the scientists who had made the discoveries, were printed upon paper fabricated from linen and cotton rags. It was not until 1786 that a book was printed upon paper in which one of the new materials was used throughout. This was a tiny book of 156 pages, printed in London, comprising the works of Charles Michel de Villette (1736–1793), and all of the text pages in one edition were printed upon paper made from the bark of the lime tree, while in another edition paper made from the marsh mallow was used. The copy before me is upon paper of the latter variety, and while the pages are a little discoloured the paper is crisp and well-preserved. At the back of

1 Proefnemingen en Monster-bladen, om Papier te Maaken Zonder Lompen, of met een gering Byvoegzel, derzelven. Amsterdam, 1770. 14 x 19 cm.

the book there are single sheet specimens of papers which had been made from nettles, hops, moss, reed grasses, three species of conferva, dandelion roots, prick wood or spindle tree bark with its pellicle and crust, hazel nut, wood of the elm, bark of the lime or linden tree, leaves of the burdock, oak, and thistle; poplar, oak, and osier bark. The paper for this book, as well as the specimens, were made by Léorier Delisle, director of the Langlée paper mills, and in a prefatory note he writes: "To Monsieur le Marquis Ducrest: If my efforts, more than my success, in the art of papermaking have earned for me a recognition on your part, and if my long travels and experiences can permit me to aspire to the protection of the august Prince, whom you represent, I owe it to your clear impartiality which characterizes your judgment. I have submitted to the paper manufacturer all of the plants, the barks, and the common vegetables, and the papers that are at the back of this volume are the results of my experiences. I wish to prove that these materials may be substituted for the usual papermaking materials, which become rarer each day. You have been the first, Monsieur, to appreciate the advantage of this discovery and I emplore you to permit me to show the public the results of my knowledge." Delisle had been preceded by Dr. Schäffer in the use of materials to replace rags for papermaking, but the specimens made by Schäffer were fabricated with an admixture of linen and cotton. Delisle, however, used nothing but the fibres of the plants in the making of his paper and he therefore claimed greater accomplishments than his predecessor. The question was carried before the Académie des Sciences and a report relating to the work of the two scientists was presented on November 18, 1786. This report was signed by Berthollet, Lavoisier et Sage and stated that the papers of Delisle contained almost no linen or cotton while those of Dr. Schäffer contained considerable rag fibre. An engraving of the Langlée mill, where Delisle carried on his experiments,

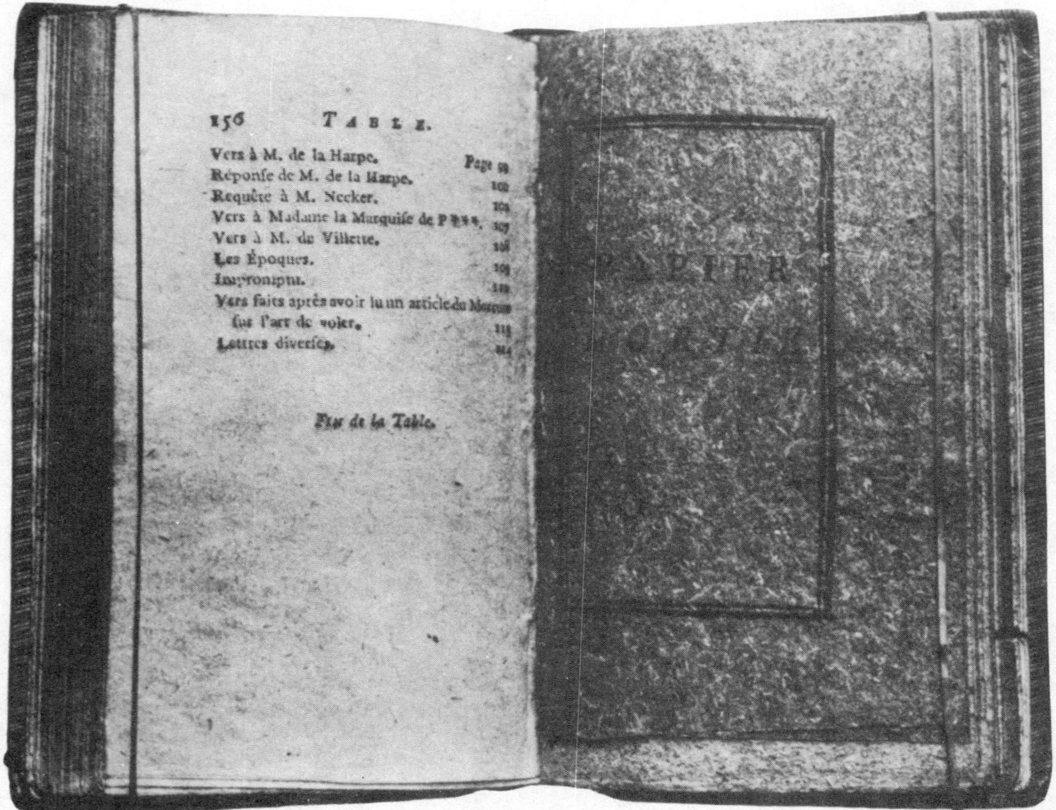

NO. 41

The little book of poems by Charles Michel de Villette which was printed on paper made from the marsh mallow. At the end of the volume there are given various specimens of paper that

NO. 42

The Langlée paper mill where Léorier Delisle made his experiments with materials for paper-making. It was in this mill that all of the paper for Charles Michel de Villette's book was made, as well as the samples showing the use of various vegetable fibres

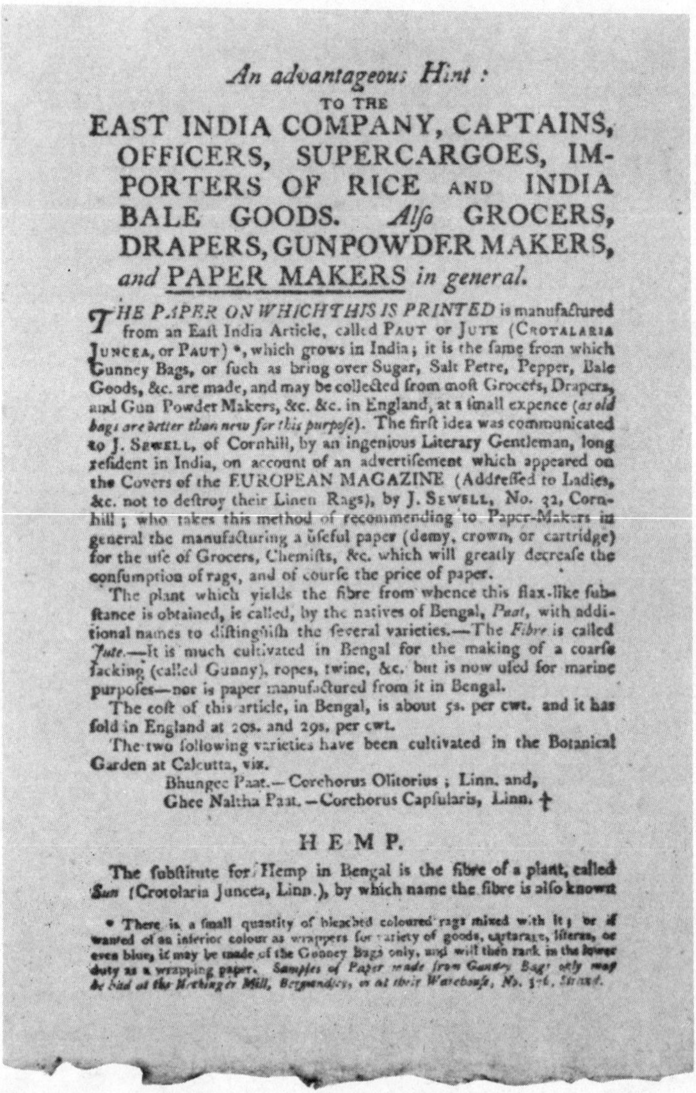

An advantageous Hint:

TO THE

EAST INDIA COMPANY, CAPTAINS, OFFICERS, SUPERCARGOES, IMPORTERS OF RICE AND INDIA BALE GOODS. *Alſo* GROCERS, DRAPERS, GUNPOWDER MAKERS, *and* PAPER MAKERS *in general.*

THE PAPER ON WHICH THIS IS PRINTED is manufactured from an Eaſt India Article, called PAUT or JUTE (CROTALARIA JUNCEA, or PAUT) *, which grows in India; it is the ſame from which Gunney Bags, or ſuch as bring over Sugar, Salt Petre, Pepper, Bale Goods, &c. are made, and may be collected from moſt Grocers, Drapers, and Gun Powder Makers, &c. &c. in England, at a ſmall expence (*as old bags are better than new for this purpoſe*). The firſt idea was communicated to J. SEWELL, of Cornhill, by an ingenious Literary Gentleman, long reſident in India, on account of an advertiſement which appeared on the Covers of the EUROPEAN MAGAZINE (Addreſſed to Ladies, &c. not to deſtroy their Linen Rags), by J. SEWELL, No. 32, Cornhill; who takes this method of recommending to Paper-Makers in general the manufacturing a uſeful paper (demy, crown, or cartridge) for the uſe of Grocers, Chemiſts, &c. which will greatly decreaſe the conſumption of rags, and of courſe the price of paper.

The plant which yields the fibre from whence this flax-like ſubſtance is obtained, is called, by the natives of Bengal, *Paat*, with additional names to diſtinguiſh the ſeveral varieties.—The *Fibre* is called *Jute*.—It is much cultivated in Bengal for the making of a coarſe ſacking (called Gunny), ropes, twine, &c. but is now uſed for marine purpoſes—nor is paper manufactured from it in Bengal.

The coſt of this article, in Bengal, is about 5s. per cwt. and it has ſold in England at 20s. and 29s. per cwt.

The two following varieties have been cultivated in the Botanical Garden at Calcutta, viz.

Bhungee Paat.—Corchorus Olitorius; Linn. and,
Ghee Naltha Paat.—Corchorus Capſularis, Linn. †

HEMP.

The ſubſtitute for Hemp in Bengal is the fibre of a plant, called *Sun* (Crotolaria Juncea, Linn.), by which name the fibre is alſo known

* There is a ſmall quantity of bleached coloured rags mixed with it; or if wanted of an inferior colour as wrappers for variety of goods, cartarage, ſiſteras, or even blue, it may be made of the Gunney Bags only, and will then rank in the lower duty as a wrapping paper. *Samples of Paper made from Gunny Bags only may be had at the Neckinger Mill, Bermondſey, or at their Warehouſe, No. 376, Strand.*

NO. 43

The first page of the pamphlet setting forth the use of hemp as a substitute for linen and cotton for papermaking forms one of the illustrations in Nicolas Desmarets' book on papermaking which is described in this chapter.

In 1797 there appeared in London a pamphlet which set forth jute as a papermaking material, the pamphlet being printed upon paper made from this substance. The sixteen page leaflet is entitled: "An Advantageous Hint: To the East India Company, Captains, Officers, Supercargoes, Importers of Rice and India Bale Goods. Also Grocers, Drapers, Gunpowder Makers, and Paper Makers in general." The pamphlet has the following to say relating to jute as a papermaking material: "The paper on which this is printed is manufactured from an East India article, called Paut or Jute (Crotalaria Juncea, or Paut), there is a small quantity of bleached coloured rags mixed with it . . . The first idea was communicated to J. Sewell, of Cornhill, by an ingenious Literary Gentleman, long resident in India, on account of an advertisement which appeared on the covers of the 'European Magazine' (Addressed to Ladies, not to destroy their Linen Rags), by J. Sewell, No. 32, Cornhill; who takes this method of recommending to Papermakers in general the manufacturing a useful paper (demy, crown, or cartridge), for use of Grocers, Chemists, &c. which will greatly decrease the consumption of rags, and of course the price of paper."

The next writer on the subject of papermaking materials to take the place of linen and cotton rags was G. A. Senger[1] who compiled what is now an exceedingly rare pamphlet. The essay is printed on paper that was made from conferva, a water plant, called by Senger "water wool." He states that this water wool, or river paper, was the oldest form of papermaking in nature. This material he termed a coralline product, being the web of water insects, seed capsules,

1 Die älteste Urkunde der Papierfabrikation in der Natur entdeckt nebst Vorschlägen zu neuen Papierstoffen von G. A. Senger, Prediger zu Reck. Dortmund and Leipzig, 1799. 13 x 19 cm.

or insects' eggs. Senger discussed the probability of sufficient conferva being found to act as a substitute for rags as a papermaking material. As Réaumur was the first writer to give account of the wasp as a fabricator of paper, Senger was the earliest scientist to dwell at length upon conferva as a form of paper, also a product of nature. Senger relates: "In my walks on the border of a small brook, I found both shores on the side of the hedges covered with a slimy substance, which the overflowed brook had deposited. The surface of the water was covered anew with a yellowish green vegetation, and in the windings of the stream there lay quantities of this fine vegetable product piled in heaps, which gave additional beauty to the blooming shores of the flowing brook. The appearance of this beauty and the thought of a useful application of this material awakened my interest, for I could not persuade myself that thrifty Nature would have brought forth so much beauty and such a great quantity of the fleecy substance, without its having a useful purpose. This covering extending over the surface of the water, was not only a resting place for insects of various sorts, and a well secured storehouse for their broods, but as Nature intends everywhere to give many advantages, I soon experienced that it contained a proper stuff for the making of paper, and what is more surprising, a paper prepared by Nature alone, without the assistance of imitating processes.

"This peculiar web contains innumerable fleecy parts of vegetation, which are generated, in the first part of the spring, on ponds and other standing waters; they detach themselves from the bottom, and rise on the surface, where they appear as a handsome green and yellow covering. After these fleecy particles have remained for some time on the watery mirror; by the heat of the sun, and by the changing degrees of cold and warmth of the water, they become more united and felted together, bleached, and at last turned into a tough paper-like covering. Or, if this fleecy substance

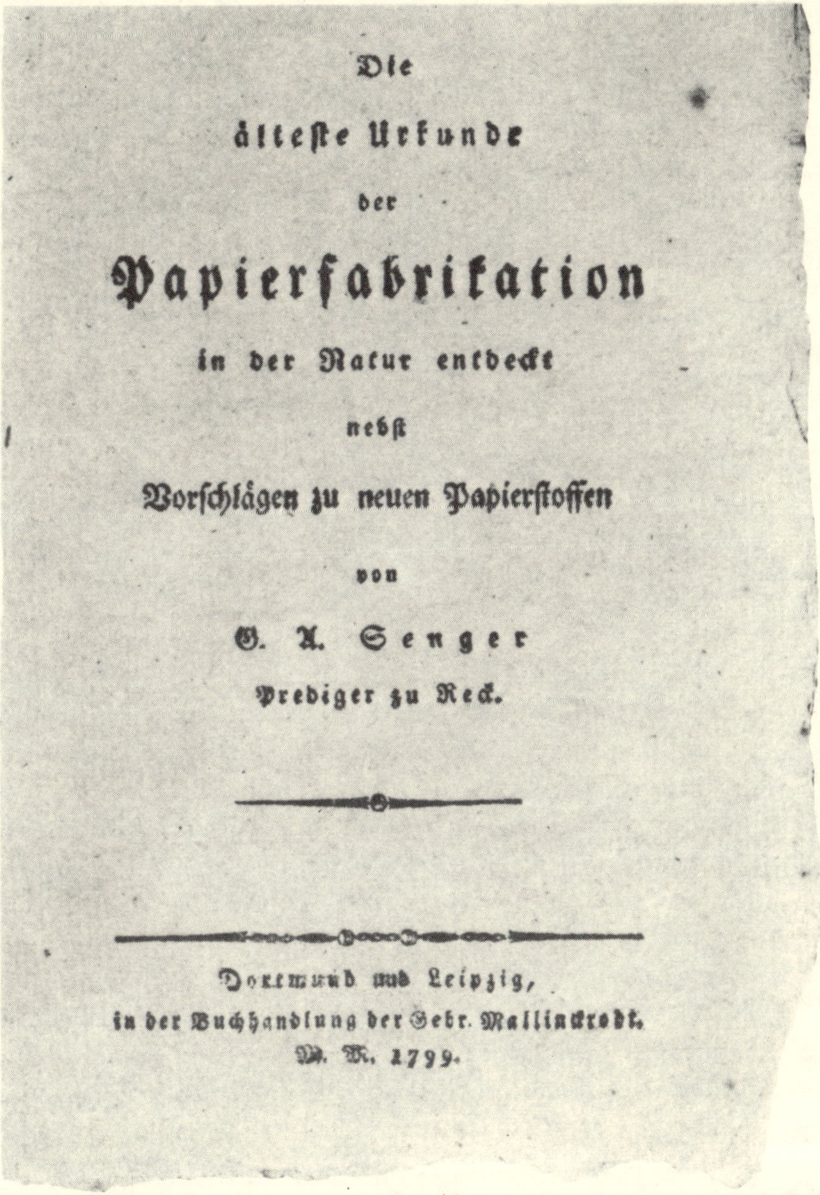
Die

älteste Urkunde

der

Papierfabrikation

in der Natur entdeckt

nebst

Vorschlägen zu neuen Papierstoffen

von

G. A. Senger

Prediger zu Reck.

Dortmund und Leipzig,

in der Buchhandlung der Gebr. Mallinckrodt.

M. M. 1799.

No. 44

Title-page of the only edition of G. A. Senger's pamphlet dealing with papermaking in nature

is mixed together, and carried away by sudden inundations, occasioned by heavy rains, and deposited on the shores, it then appears like a thin jelly or slime, which, after it has undergone several changes naturally produced by the contents of air and water, turns

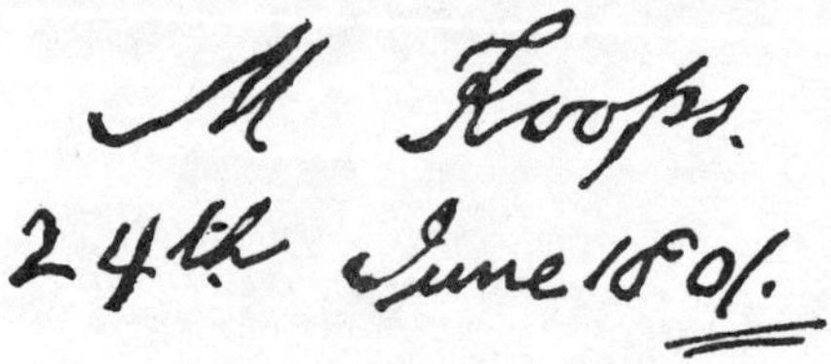

NO. 45
Matthias Koops' signature, 1801

into a kind of paper, which resembles the common paper; or, where it has been produced upon clear water, it is not unlike a superior paper, of which some may be gathered nearly as white as writing paper." The conferva paper used in Senger's book is of a green-grey colour, but even after one hundred and thirty years, the paper is stout and in as good condition as when it was made; there was, however, no chemical bleaching used in its preparation. Senger's pamphlet consists of ninety-six pages with a paper cover.

In the course of searching for new papermaking materials, we now come to the work of Matthias Koops who published his first book on this subject in London in 1800.[1] The earliest edition of this notable work was "printed on the first useful paper manufactured solely from straw." The paper is of the "laid" type and bears the crown and G R watermark. There is an appendix of seven pages "printed on paper made from wood alone, the product of this country, without any intermixture of rags, waste paper, bark,

1 Historical account of the substances which have been used to describe events, and to convey ideas, from the earliest date, to the invention of paper. London, 1800. 16 x 27 cm.

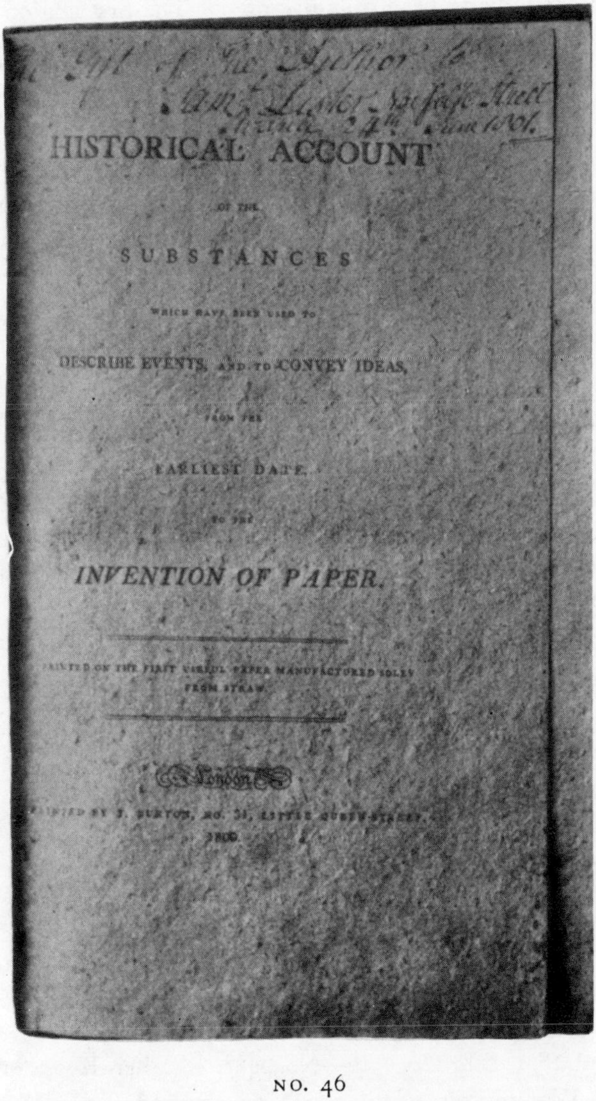
HISTORICAL ACCOUNT

OF THE

SUBSTANCES

WHICH HAVE BEEN USED TO

DESCRIBE EVENTS, AND TO CONVEY IDEAS,

FROM THE

EARLIEST DATE,

TO THE

INVENTION OF PAPER.

PRINTED ON THE FIRST USEFUL PAPER MANUFACTURED SOLELY FROM STRAW.

London:

NO. 46

Title-page of Matthias Koops' book which was "printed on the first useful paper manufactured solely from straw"

straw, or any other vegetable substance, from which paper might be, or has hitherto been manufactured." The first edition of this book embraces eighty-two pages of text. The second edition of Matthias Koops' work (London, 1801. 12 x 20 cm.), was "printed on paper re-made from old printed and written paper," with a sixteen page appendix on paper made from wood. The paper used in this edition is of the "wove" type without watermarks. Another printing of this edition, the rarest of the three, was issued in 1801, and unlike the regular second edition, was printed on paper made from straw. Only a limited number of these copies were published. An engraving (27 x 36 cm.), entitled: "Noon" by Austin, after J. Ruysdael appeared at the beginning of the nineteenth century with the caption: "On paper made from straw alone, by Mr. Koops, 1800."

Mr. Koops' experiments in the use of straw and wood for papermaking were carried on at Neckinger mill, Bermondsey, and the three following patents pertaining to his researches are recorded in the British Great Seal Patent office:

"A.D. 1800, April 28th—No. 2392. Matthias Koops of Queen Street, Ranelagh, in the county of Middlesex, gentleman, for a mode of extracting printing and writing ink from printed and written paper, and converting the paper from which the ink is extracted into pulp, and making thereof paper fit for writing, printing, and other purposes."

"A.D. 1800, August 2nd—No. 2433. Matthias Koops of Queen Street, Ranelagh, in the county of Middlesex, gentleman, for a method of manufacturing paper from straw, hay, thistles, waste, and refuse of hemp and flax, and different kinds of wood and bark."

"A.D. 1801, February 17th—No. 2481. Matthias Koops of James Street, Westminster, gentleman, for a method of manufacturing paper from straw, hay, thistles, waste, and refuse of hemp and flax,

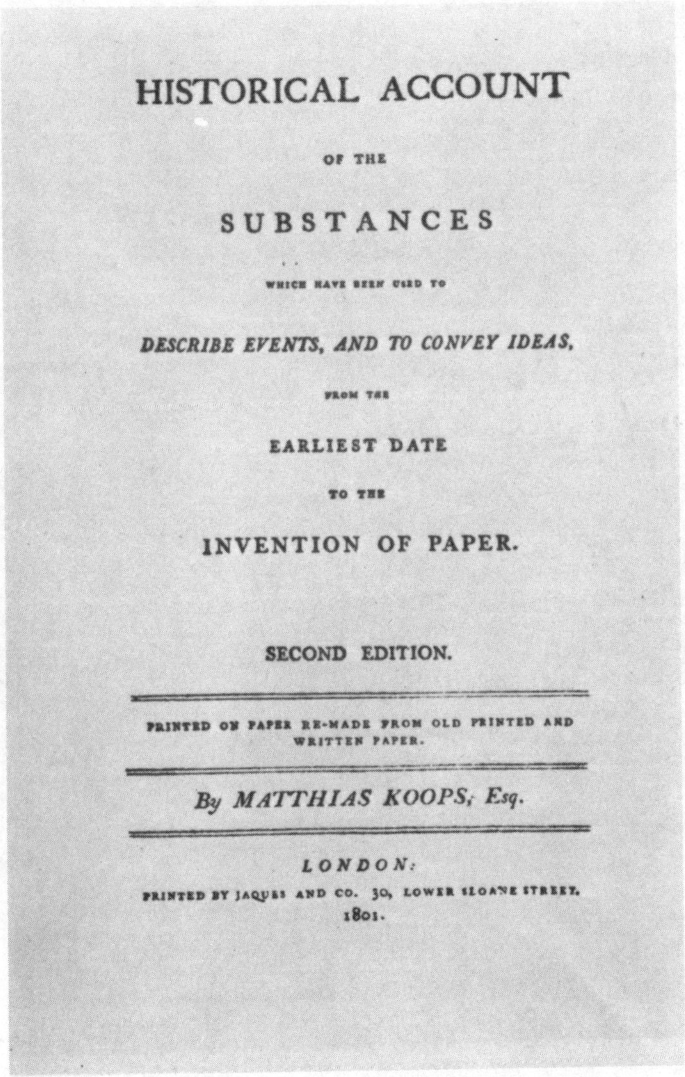

HISTORICAL ACCOUNT

OF THE

SUBSTANCES

WHICH HAVE BEEN USED TO

DESCRIBE EVENTS, AND TO CONVEY IDEAS,

FROM THE

EARLIEST DATE

TO THE

INVENTION OF PAPER.

SECOND EDITION.

PRINTED ON PAPER RE-MADE FROM OLD PRINTED AND WRITTEN PAPER.

By MATTHIAS KOOPS, Esq.

LONDON:

PRINTED BY JAQUES AND CO. 30, LOWER SLOANE STREET.
1801.

NO. 47

Title-page of the second edition of Koops' book, "printed on paper re-made from old printed and written paper," with an appendix made from wood and different kinds of wood and bark, fit for printing and other useful purposes."

The three editions of Koops' work are practically identical in text and treat chiefly of the varied materials used in past ages as

NO. 48

A Fabriano paper mill as depicted in *Novum Italiae Theatrum,* 1724

substances upon which to write. Although Matthias Koops' work is comparatively recent, little is known of the man, and a diligent search in the British Museum and Patent Office Libraries reveals nothing concerning him.

Koops' book completes the list of works dealing with the subject of papermaking materials, which appeared up to the beginning of the nineteenth century, beyond which time it is not my purpose to bring the descriptions in this chapter. It will now be necessary to go back to the point where I deviated from chronological order and give some account of the books and pamphlets which relate to the history and actual making of paper.

In an early eighteenth century Italian work[1] unrelated to the art of papermaking, there may be found an interesting engraving depicting the interior of a paper mill. The folio volumes of this describe the cities, palaces, and sacred buildings of Italy, and in the second volume, plate xxvii, the view of a Fabriano mill may be studied. Illustrations of papermaking in books foreign to the subject are rare.

After the great technical volume of Leonhardt Christoph Sturm, described in Part III, no important work of this kind was issued until 1734, when there were published at Amsterdam the two folio volumes compiled by Leendert van Natrus. This book, "Groot Volkomen Moolenboek"[2] is a rare treatise on machinery and the equipment of windmills. Plates 16, 17, 18, and 19 relate to paper manufacture and two text pages are devoted to this subject. The first plate outlines the general floor plan of a paper mill, with the arrangement of the various appliances used in the craft. In the second plate, devoted to papermaking, the construction of the presses is shown, also the method used in hanging the hair ropes upon which paper was dried. Plate 18 gives engravings of the "Hollanders," or beating engines, which were used for the preparation of papermaking materials, the first depicting of these machines in

1 Novum Italiae Theatrum sive accurata Descriptio ipsius urbium, palatiorum, sacrarum aedium, etc. Hagae-comitum, 1724. 4 volumes. 35 x 47 cm.

2 Groot Volkomen Moolenboek; of, Naauwkeurig Ontwerp Van Allerhande tot nog toe bekende Soorten van Moolens, Met haare Gronden en Opstallen, en al het geene verder daar toe behoort. Ten dienste der Oeffenaaren van deze Kunst, en van alle Liefhebbers der Bouwen Tekenkunst, geteekent door Leendert Van Natrus, Moolemakers Baas van de Ed: Oost-Indische Compagnie, te Amsterdam, Jacob Polly, Moolemaker te Saardam; en Cornelis Van Vuuren; En zeer naauwkeurig in het Kooper gebragt door Jan Punt. Amsterdam, 1734–35. 2 volumes in one. 33 x 54 cm.

GROOT VOLKOMEN
MOOLENBOEK;
OF
NAAUWKEURIG ONTWERP
Van Allerhande tot nog toe bekende Soorten van
MOOLENS,
Met haare GRONDEN en OPSTALLEN,
en al het geene verder daar toe behoort.
Ten dienste der Oeffenaaren van deze Kunst, en van alle Liefhebbers der Bouw- en Tekenkunst, geteekent
DOOR
LEENDERT VAN NATRUS,
Moolenmakers Baas van de Ed: Oost-Indische Compagnie, te Amsterdam,
JACOB POLLY,
Moolemaker te Saardam;
EN
CORNELIS VAN VUUREN,
En zeer nauwkeurig in het Kooper gebragt
Door JAN PUNT.
EERSTE DEEL

Te AMSTERDAM,
By JOHANNES CÓVENS EN CORNELIS MORTIER.
MDCCXXXIV.

NO. 49

Title-page of the Dutch book on mill machinery in which the first engraving of the Hollander appears in the country of its invention

the country of their invention. In plate 19 is shown a windmill driving the Hollanders, arranged in such positions that the greatest quantity of rags could be beaten with the least amount of power.

NO. 50

John Lewis the first writer in English to touch on the subject of watermarks

The first writer in English to touch on the subject of watermarks was John Lewis (1675–1747), who compiled his history of printing[1] in 1737. Lewis considered papermarks of sufficient importance to devote two pages to reproductions of about twenty marks which appear in the paper upon which, from 1477 to 1490 Caxton

1 The Life of Mayster Wyllyam Caxton, of the Weald of Kent; the First Printer in England. In which is given An Account of the Rise and Progress of the Art of Pryntying in England, during his Time, till 1493. London, 1737. 13 x 22 cm.

books had been printed. The marks include such devices as shields, animals, shears, bull's heads; also a crown, a hand, a ship, and the letter "P," which is common in Caxton books, probably the initial of Philip the Good. In German the earliest book that touched

NO. 51

Reproductions of the watermarks found in the paper used by William Caxton. From the book of 1737 on typography by John Lewis who was the first English writer to regard watermarks

upon watermarks was compiled by Christian Gottlieb Schwarz[1] and published in 1740. In this volume, a few papermarks are reproduced, which constitutes the only papermaking interest in the book. In 1749 Joseph Ames[2] (1688–1758), in his history of print-

1 Primaria qvaedam docvmenta de Origine Typographiae qvorvm illvstratorvm partem primam svb praesidio Christiani Gottlibii Schwarzii. Altorfii, 1740. 15 x 17 cm.

2 Typographical Antiquities: Being an Historical Account of Printing in England: with some Memoirs of our Ancient Printers, and a Register of Books printed by them, from the Year mccclxxi to the Year mdc. London, 1749. 19 x 26 cm.

ing in England considered watermarks of sufficient significance to devote two pages of reproductions to them. The marks chosen by Ames are the same as those depicted by John Lewis about ten years earlier. Both Lewis and Ames term these devices "papermarks" and not "watermarks," an appellation not in use in English until about 1790.

The first book of trades in the English language which embraced the art of papermaking appeared in 1747,[1] and was founded on the much earlier works of the same kind in German and Dutch. These books were compiled for the benefit of young men in search of a craft or calling, and gave an outline of each occupation, the length of the apprenticeship, the daily hours of work, and above all, the requirements of the master who operated the mill or shop. Toward the close of the eighteenth century these little books of trades became quite common, in fact, so common that I shall not enumerate them; there are probably fifty such items in my collection.

One of the important Dutch books which deal with papermaking was issued in Amsterdam in 1761[2] and was founded on the earlier work of Leendert van Natrus. The author of this later work, Johannis van Zyl, was an authority on windmill construction, and most of the volume is devoted to that subject. There are, however, six engravings relative to the art of papermaking, with four pages of explanatory text, giving a complete account of the craft as it was practiced at that time. In this same year (1761), the most complete treatise on the art of papermaking which had up to

1 A general description of all Trades, digested in alphabetical order; by which Parents, Guardians, and Trustees, may, with greater ease and certainty, make choice of Trades agreeable to the Capacity, Education, Inclination, Strength, and Fortune of the Youth under their Care. London, 1747. 10 x 17 cm.

2 Theatrum Machinarum Universale; of Groot Algemeen Moolen-Boek, Behelzende de Beschryving en Afbeeldingen van allerhande soorten van Moolens, . . . En in't Koper gebragt door Jan Schenk. Amsterdam, 1761. 38 x 54 cm.

THEATRUM
MACHINARUM UNIVERSALE;
OF
GROOT ALGEMEEN
MOOLEN-BOEK,

Behelzende de BESCHRYVING en AFBEELDINGEN van allerhande ſoorten van MOOLENS, der zelver OPSTALLEN, en GRONDEN;

Waar onder veele, die eerſt onlangs zyn gebouwt, ſtaande zo buiten de Stad Amſterdam, als in de omleggende Plaatſen; nevens eene naauwkeurige Verbeelding in verſcheide Platen van de AMSTERDAMSCHE MODDER-MOOLENS; ook een nette Verdeeling van de Staven en Kammen, en meer andere zeer fraaie Werken ten nutte der BOUW- en TEKEN-KONST; nooit zo volkomen in 't licht gebragt.

Voor 't grootſte gedeelte getekent te Amſterdam

DOOR

JOHANNIS VAN ZYL,

Moolenmaker van Lexmondt;

En in 't Koper gebragt door

JAN SCHENK.

I. DEEL.

Te AMSTERDAM,
By PETRUS SCHENK, Konſt- en Kaartverkoper, in de Kalverſtraat, by de Gaperſteeg, in N. Viſſchers Atlas. MDCCLXI.

NO. 52
Title-page of the work by the Dutch scientist, Johannis van Zyl

NO. 53
Joseph Jérôme Le Français de Lalande

that date been attempted was compiled by Joseph Jérôme Le Français de Lalande (1732–1807), and published in Paris.[1] In this book there are fourteen large engravings, one folded, showing the entire process of making paper, from the sorting and cutting of the linen and cotton rags to the wrapping and packing of the finished paper. One hundred and fifty pages of text are devoted to a description of

1 Art de faire le papier. In: Description des arts et métiers, published by the Académie des sciences. Volume 4. Paris, 1761. 29 x 45 cm.

NO. 54
Nicolas Desmarets

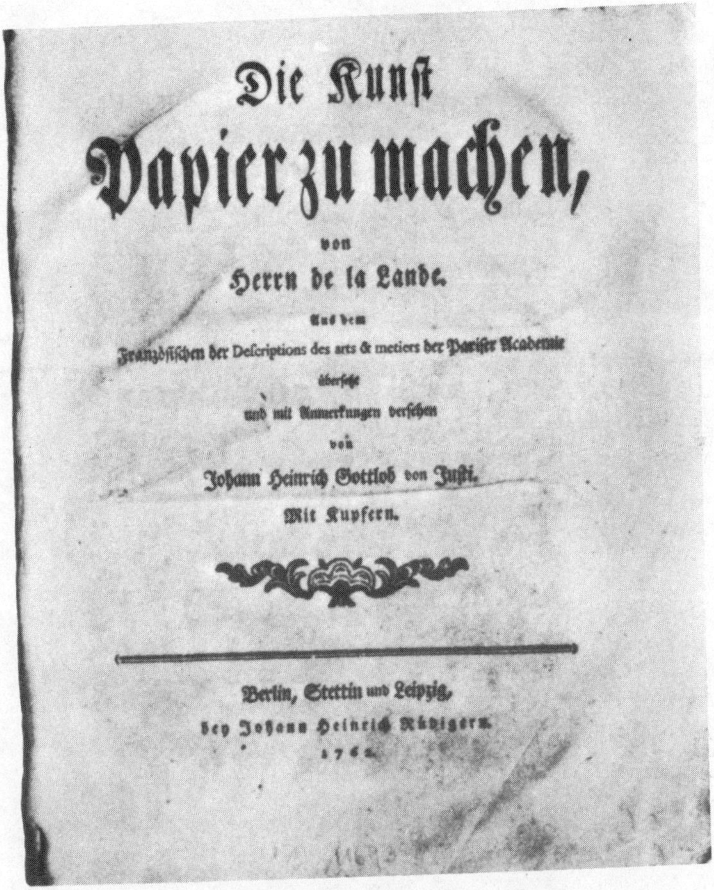

Die Kunst
Papier zu machen,
von
Herrn de la Lande.
Aus dem
Französischen der Descriptions des arts & metiers der Pariser Academie
übersetzt
und mit Anmerkungen versehen
von
Johann Heinrich Gottlob von Justi.
Mit Kupfern.

Berlin, Stettin und Leipzig,
bey Johann Heinrich Rüdigern.
1762.

NO. 55
Title-page of the German edition of de Lalande's work

the plates and to the art of papermaking in general. No other important technical work was issued in Europe until that of Nicolas Desmarets[1] (1725–1815) made its appearance in 1774. In the fif-

1 Premier Mémoire sur les Principales manipulations, Qui sont en usage dans les Papeteries de Hollande, avec l'Explication physique des résultats de ces manipulations. Lû à l'Académie Royale des Sciences, le 20 Février 1771. Par M. Desmarets, de la même Académie. Paris, 1774. 20 x 27 cm.

VOLLEDIGE
BESCHRIJVING
VAN ALLE
KONSTEN, AMBACHTEN, HANDWERKEN, FABRIEKEN, TRAFIEKEN, DERZELVER WERKHUIZEN, GEREEDSCHAPPEN, ENZ.
TEN DEELE OVERGENOMEN
UIT DE BEROEMDSTE BUITENLANDSCHE
WERKEN;
En vermeerderd met de Theorie en Praktijk der beste Inlandsche Konstenaaren en Handwerkslieden.

NEGENDE STUK.

DE
PAPIERMAAKER.

MET PLAATEN.

Te DORDRECHT,
BIJ A. BLUSSE EN ZOON.
MDCCXCII.

NO. 56
Title-page of the Dutch edition of Joseph Jérôme Le Français de Lalande's great work on papermaking

teen finely-engraved plates of this book there is given a complete delineation of papermaking. It was from the works of Leendert van Natrus, Johannis van Zyl, Joseph de Lalande, and N. Desmarets that the papermakers of the eighteenth century gained their

NO. 57
Gerard Meerman

knowledge, and very fine text-books they were too, for even today it is possible to learn from these volumes a great deal regarding the making of handmade paper. A Dutch edition[1] of the de Lalande work was issued in quarto in Dordrecht in 1792. In this edition the engravings are identical with those of the parent edition, ex-

1 Volledige beschrijving van alle konsten, ambachten, handwerken, fabrieken, trafieken, derzelver werkhuizen, gereedschappen, enz. ten deele overgenomen uit de beroemdste buitenlandsche werken . . . de papiermaaker. Dordrecht, 1792. 13 x 23 cm.

GERARDI MEERMAN

ET DOCTORUM VIRORUM
AD EUM EPISTOLAE
ATQUE OBSERVATIONES

DE

CHARTAE VULGARIS SEU
LINEAE ORIGINE.

EDIDIT

AC PRAEFATIONE INSTRUXIT

JACOBUS VAN VAASSEN.

HAGAE-COMITUM,
Apud NICOLAUM VAN DAALEN.
MDCCLXVII.

NO. 58
Title-page of the second compilation by Gerard Meerman on the origin of linen paper

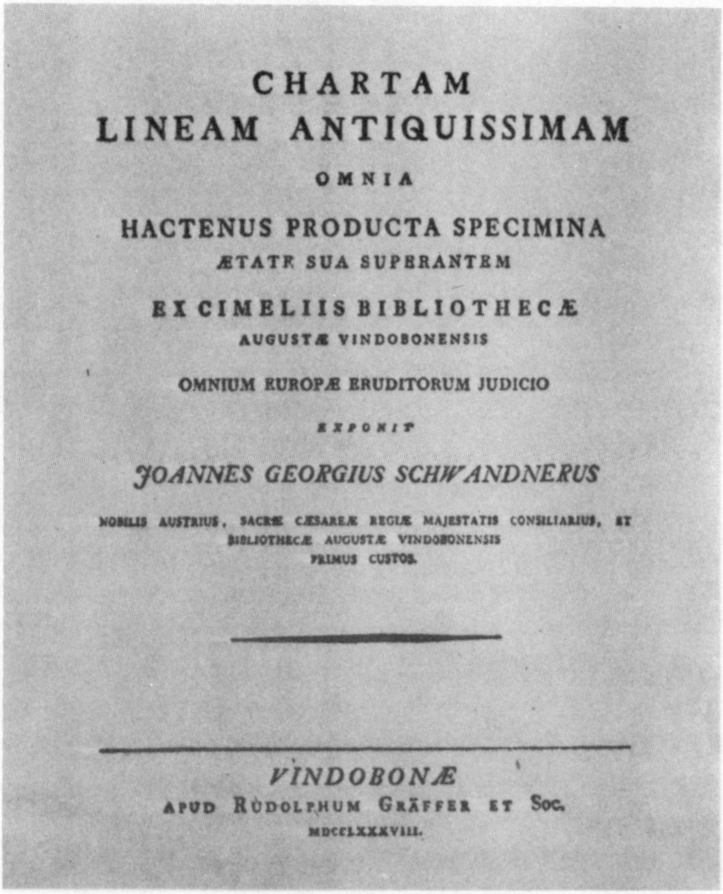

CHARTAM
LINEAM ANTIQUISSIMAM
OMNIA
HACTENUS PRODUCTA SPECIMINA
ÆTATE SUA SUPERANTEM
EX CIMELIIS BIBLIOTHECÆ
AUGUSTÆ VINDOBONENSIS
OMNIUM EUROPÆ ERUDITORUM JUDICIO
EXPONIT
JOANNES GEORGIUS SCHWANDNERUS
NOBILIS AUSTRIUS, SACRÆ CÆSAREÆ REGIÆ MAJESTATIS CONSILIARIUS, ET BIBLIOTHECÆ AUGUSTÆ VINDOBONENSIS PRIMUS CUSTOS.

VINDOBONÆ
APUD RUDOLPHUM GRÄFFER ET SOC.
MDCCLXXXVIII.

NO. 59

Title-page of the work by Johann Georg Schwandtner dealing with the earliest linen paper

cept for slight modifications in the costumes of the workers. An edition in Spanish[1] of Lalande's work was also thought necessary, this was published in Madrid and appeared in 1778.

1 Arte de hacer el papel según se practica en Francia, y Holanda, en la China, y en el Japón. Traduc. por. Mig. Geron. Suarez y Nuñez, Madrid, 1778.

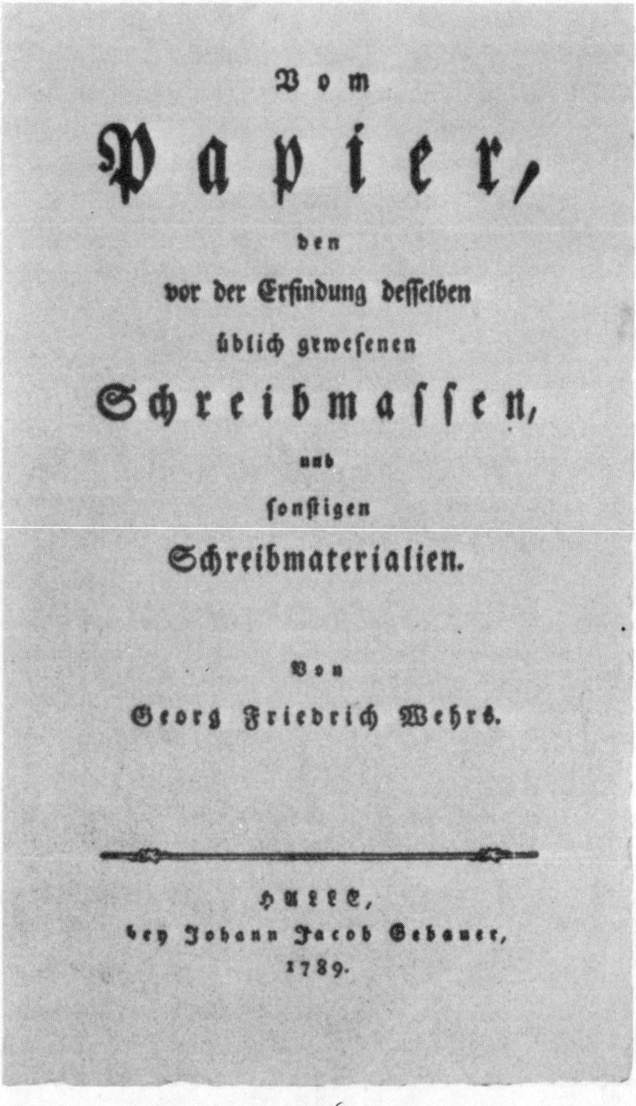

Vom

Papier,

den

vor der Erfindung desselben

üblich gewesenen

Schreibmassen,

und

sonstigen

Schreibmaterialien.

Von

Georg Friedrich Wehrs.

Halle,

bey Johann Jacob Gebauer,

1789.

NO. 60

Title-page of the original edition of Georg Friedrich Wehrs' book on the history of paper

Aside from these technical books there were numerous works throughout the eighteenth century that dealt with early writing appliances and the history of the use of linen and cotton as paper-making materials. In 1762 Gerard Meerman (1722–1771) in his book[1] offered a premium of twenty-five ducats for the discovery of the date when linen paper was first fabricated. From all parts of Europe specimens were sent to him which were supposed to be early linen paper, but his researches established no definite conclusion regarding the origin of linen as a material for paper. In 1767 Meerman issued his second compilation[2] on the origin of linen paper. In 1788 Johann Georg Schwandtner[3] wrote his observations relating to the earliest linen paper which had up to that time been discovered. The sheet of paper in question had been found among the charters of the ancient Monastery of Göss in Upper Stiria, and was thought to date from about the year 1243. Schwandtner was Principal Keeper of the Imperial Library, Vienna, where his book was published. Following closely upon the work of Schwandtner was the comprehensive volume by Georg Friedrich Wehrs[4] (1750–1818), which remains to this day one of the greatest works on the history of early writing materials, including papyrus, parchment, and paper, that has ever been compiled. During the year 1790 a supplement to this book was written by Wehrs and published in Hanover. In 1784–1801 the two volume work of

1 Syndici Roterodamensis, Admonitio de chartae nostratis, seu lineae, origine. Rotterdam, 1762.

2 Meerman et doctorum virorum ad eum epistolae atque observationes de chartae vulgaris seu lineae origine. Hagae-Comitum, 1767. 11 x 18 cm.

3 Chartam lineam antiquissimam omnia hactenus producta specimina aetate sua superantem ex cimeliis bibliothecae Augustae Vindobonensis omnium Europae eruditorum judicio exponit Joannes Georgius Schwandnerus. Vindobonae, 1788. 18 x 24 cm.

4 Vom Papier, den vor der Erfindung desselben üblich gewesenen Schreibmassen und sonstigen Schreibmaterialien. Halle, 1789. 12 x 21 cm.

Johann Gottlob Immanuel Breitkopf[1] (1719–1794), was issued in Leipzig; while the first volume of this work gives a description of linen paper, most of the text is devoted to the history of playing cards and wood-engraving. There are several plates of interest, in-

NO. 61
Johann Gottlob Immanuel Breitkopf

cluding the interior of a Japanese paper mill, and the implements of a Hindustani paper manufactory; also engravings of old stamping machines and reproductions of numerous watermarks. Breitkopf was a celebrated printer and publisher of Leipzig, and in 1750

1 Versuch, den Ursprung der Spielkarten, die Einführung des Leinenpapieres, und den Anfang der Holtzschneidekunst in Europa zu erforschen. Leipzig, 1784–1801. 2 volumes. 19 x 24 cm.

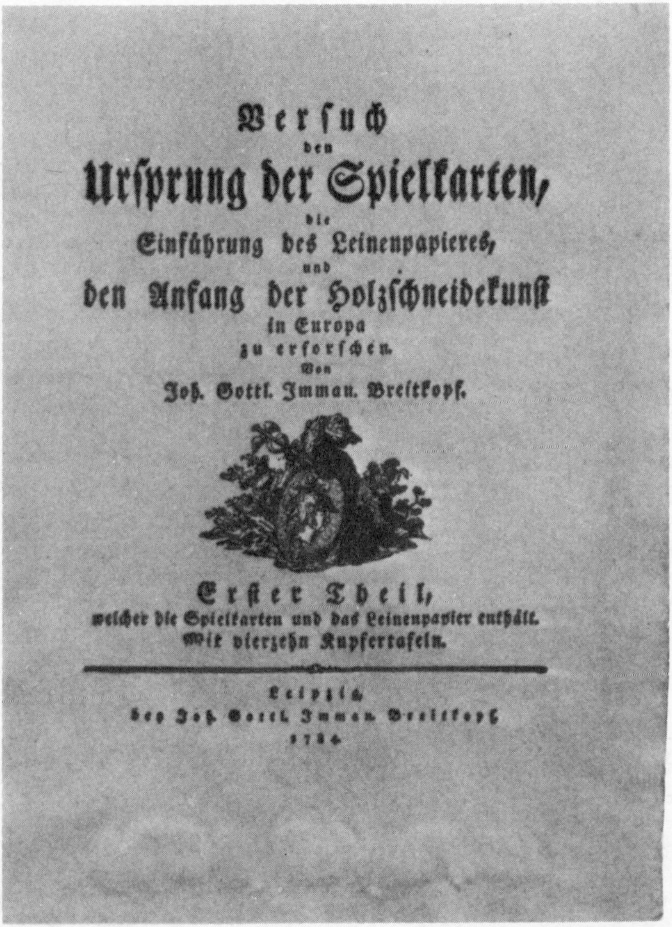

Versuch
den
Ursprung der Spielkarten,
die
Einführung des Leinenpapieres,
und
den Anfang der Holzschneidekunst
in Europa
zu erforschen.
Von
Joh. Gottl. Imman. Breitkopf.

Erster Theil,
welcher die Spielkarten und das Leinenpapier enthält.
Mit vierzehn Kupfertafeln.

Leipzig,
bey Joh. Gottl. Imman. Breitkopf.
1784

NO. 62

Title-page of the two volume work by Breitkopf which deals with linen paper

invented movable types for printing music, he being a musician of considerable ability.

The process of bleaching linen and cotton rags was invented in 1774 by Karl Wilhelm Scheele (1742–1786), a celebrated German

NO. 63

Karl Wilhelm Scheele, the inventor of the process of bleaching linen and cotton rags for papermaking

chemist, living in Sweden. The earliest work in English on the bleaching of paper that has come to my attention is dated 1792[1] and is a small octavo pamphlet pointing out the absurdity of one firm having the monopoly of the process. Apparently the Messrs. Taylors, English papermakers, thought that their mill should, for some reason not stated, have the exclusive use of this method of whitening their paper, and the little booklet of forty-three pages was printed in Edinburgh as a protest against this mill having an exclusive privilege. The first technical work in English relative to the chemical bleaching of paper, that can be traced, was published in London in 1799[2] and is a translation by William Nicholson of Newman Street, London, of the work of Pajot des Charmes. This book gives short articles on the method of, "bleaching written or printed papers and rags, whether unbleached, dyed or coloured"; also of the "bleaching of old written papers, to be worked up again." Inasmuch as I have never been particularly interested in the bleaching of paper, since it destroys the fibres of the sheets, the books dealing with this phase of papermaking have not especially appealed to me, although every collection of books of paper interest should, of course, include them.

Toward the latter part of the eighteenth century there were numerous books published in which paper and papermaking were mentioned. Books of this description would include the works of Gottlieb Christoph Von Murr[3] (1733–1811), Paul Von Stetten[4]

1 Memorial relative to the invention of a new method of bleaching, showing the absurdity of any pretentions to an exclusive privilege for using it in the paper manufacture. Edinburgh, 1792. 13 x 20 cm.

2 The Art of Bleaching Piece-Goods, Cottons, and Threads, of every description . . . composed for the use of manufacturers, bleachers, dyers, callico printers, and paper-makers. London, 1799. 15 x 22 cm.

3 Journal zur Kunstgeschichte und zur allgemeinen Litteratur. Nürnberg, 1777–1784. Volumes 5 and 13. 10 x 17 cm.

4 Kunstgewerb und Handwerks Geschichte der Reichs-Stadt Augsburg. Augsburg, 1779. 10 x 17 cm.

(1731–1808), Firmin Didot[1] (1764–1836), Thomas Astle[2] (1735–1803), Samuel Denne[3] (1730–1799), Henry Lemoine[4] (1756–1812), and Gaston Armand Camus[5] (1740–1804). While these books and pamphlets are insignificant, both technically and historically, they should be included in the collection of the bibliophile.

The earliest edition of the "Visible World" by Johann Amos Comenius[6] (1592–1670) in my collection is dated 1777, the twelfth edition. The first edition of this volume was the earliest pictorial school-book issued, and, like the numerous subsequent editions, attempted to explain the "chief things that are in the world, and of men's employments." The text of the first edition was written in Latin and High Dutch, "every where one word answer'd to another over against it." Papermaking is illustrated by the quaint woodcut shown in No. 75. The description refers to the numbers on the picture and reads: "The ancients used beech-boards, 1, or leaves, 2, also bark of trees, 3; especially of an Egyptian shrub which was called papyrus. Now paper is in use, which the paper-maker maketh in a paper-mill, 4, of linen rags, 5, stamped to a mash, 6, which being taken up in frames (moulds), 7, he spreadeth into sheets, 8, and setteth them in the air so that they may be dried. Twenty-five of these make a quire, 9, twenty quires a ream, 10,

1 Lettre sur les découvertes de M. Didot, aîné, dans les arts de la papeterie. Paris, 1783. (16-page pamphlet.)

2 The Origin and Progress of Writing . . . also some account of the Origin and Progress of Printing. London, 1784. 22 x 29 cm.

3 Observations on Papers. In: Archaeologia. Volume 12. London, 1795.

4 Typographical Antiquities. History, Origin, and Progress of the Art of Printing . . . with a Curious Dissertation on the Origin and Use of Paper. London, 1797. 10 x 17 cm.

5 Notice d'un livre imprimé a Bamberg en mccccxlii, lue a l'Institut National. Paris, 1799. 30 x 23 cm.

6 Visible World: or, A Nomenclature, and Pictures, of all the Chief Things that are in the World, and of Men's Employments therein; In above 150 cuts. Written by the author in Latin and High Dutch. Translated into English by Charles Hoole, M.A. London, 1777. 11 x 18 cm.

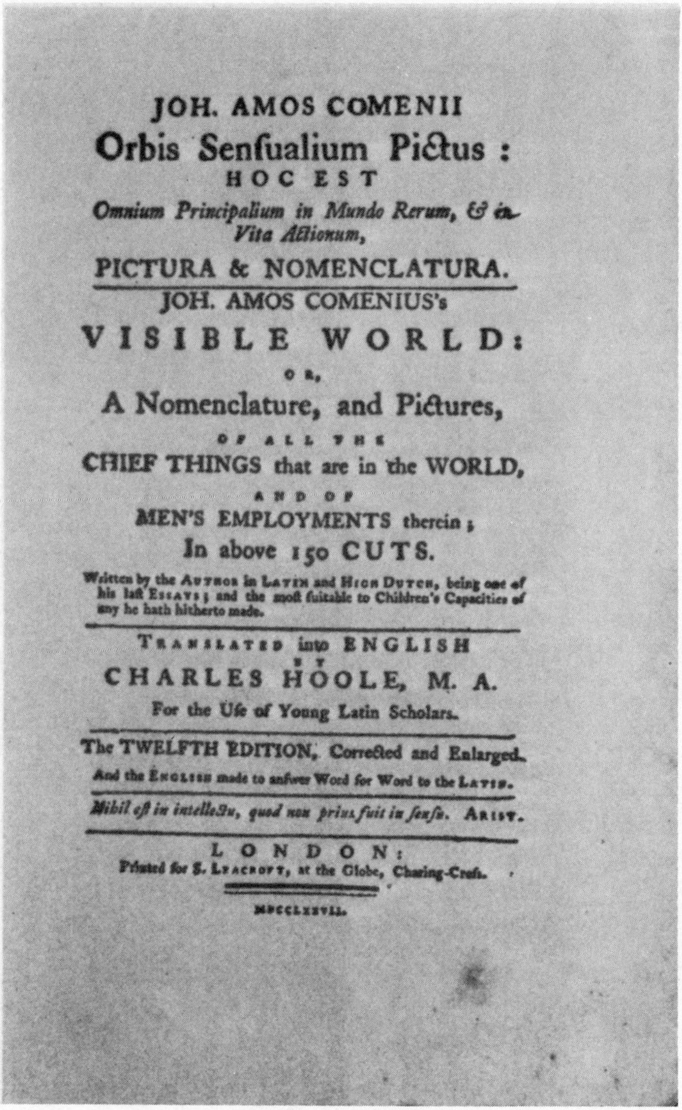

JOH. AMOS COMENII

Orbis Senſualium Pictus:

HOC EST

Omnium Principalium in Mundo Rerum, & in Vita Actionum,

PICTURA & NOMENCLATURA.

JOH. AMOS COMENIUS's

VISIBLE WORLD:

OR,

A Nomenclature, and Pictures,

OF ALL THE

CHIEF THINGS that are in the WORLD,

AND OF

MEN'S EMPLOYMENTS therein;

In above 150 CUTS.

Written by the AUTHOR in LATIN and HIGH DUTCH, being one of his laſt ESSAYS; and the moſt ſuitable to Children's Capacities of any he hath hitherto made.

TRANSLATED into ENGLISH

BY

CHARLES HOOLE, M. A.

For the Uſe of Young Latin Scholars.

The TWELFTH EDITION, Corrected and Enlarged.

And the ENGLISH made to anſwer Word for Word to the LATIN.

Nihil eſt in intellectu, quod non prius fuit in ſenſu. ARIST.

LONDON:

Printed for S. LEACROFT, at the Globe, Charing-Croſs.

MDCCLXXVII.

NO. 64

Title-page of the twelfth edition of the *Visible World* by Johann Amos Comenius

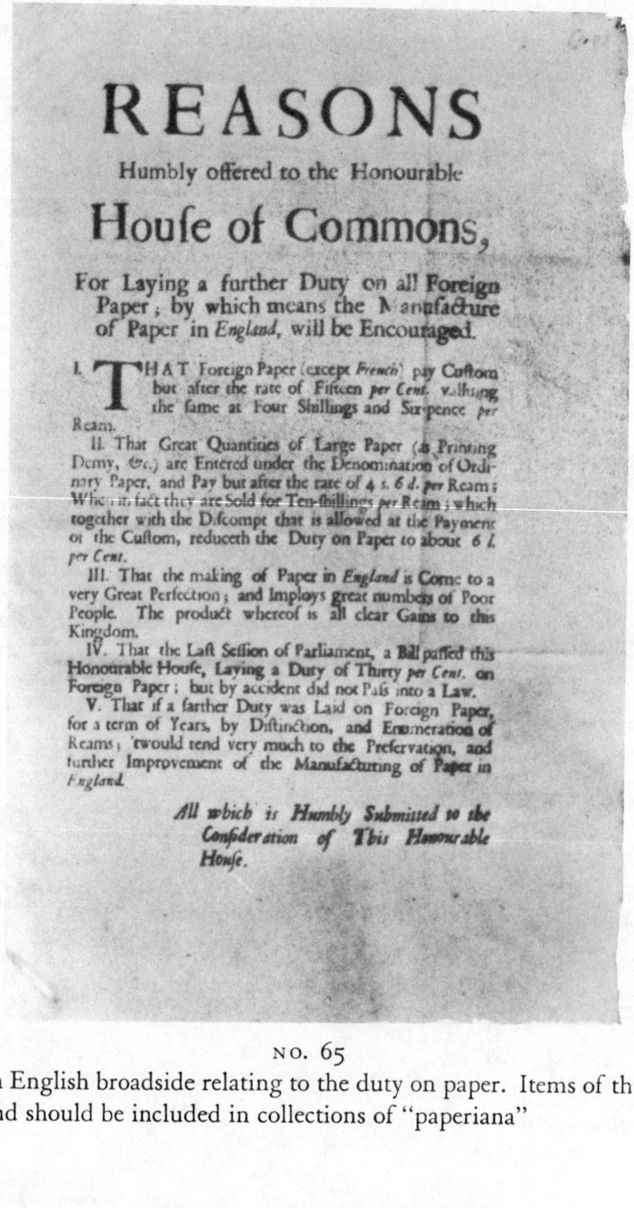

REASONS

Humbly offered to the Honourable

Houſe of Commons,

For Laying a further Duty on all Foreign Paper; by which means the Manufacture of Paper in *England*, will be Encouraged.

I. THAT Foreign Paper (except *French*) pay Cuſtom but after the rate of Fifteen *per Cent.* valluing the ſame at Four Shillings and Six-pence *per* Ream.

II. That Great Quantities of Large Paper (as Printing Demy, *&c.*) are Entered under the Denomination of Ordinary Paper, and Pay but after the rate of 4 *s.* 6 *d. per* Ream; When in fact they are Sold for Ten-ſhillings *per* Ream; which together with the Diſcompt that is allowed at the Payment of the Cuſtom, reduceth the Duty on Paper to about 6 *l. per Cent.*

III. That the making of Paper in *England* is Come to a very Great Perfection; and Imploys great numbers of Poor People. The product whereof is all clear Gains to this Kingdom.

IV. That the Laſt Seſſion of Parliament, a Bill paſſed this Honourable Houſe, Laying a Duty of Thirty *per Cent.* on Foreign Paper; but by accident did not Paſs into a Law.

V. That if a farther Duty was Laid on Foreign Paper, for a term of Years, by Diſtinction, and Enumeration of Reams; 'twould tend very much to the Preſervation, and further Improvement of the Manufacturing of Paper in *England*.

All which is Humbly Submitted to the Conſideration of This Honourable Houſe.

NO. 65

An English broadside relating to the duty on paper. Items of this kind should be included in collections of "paperiana"

and ten of these a bale of paper, 11. That which is to last long is written on parchment, 12." Another little book of the same description as the "Visible World" was published in London in 1791, under the general title of "The Progress of Man and Society." This little book of trades was compiled by the Reverend Doctor Trusler and its mission was to instruct the young in the choice of a calling or profession. The chief papermaking interest in this book lies in the wood engraving reproduced in illustration No. 94, which is not found in any other book of trades, either English or continental. This same illustration, however, may be found on a coloured broadside which was published in London in 1817 with the caption: "He that hath a Calling, hath an Estate."

On June 20, 1788, Benjamin Franklin read before the members of the American Philosophical Society, Philadelphia, a treatise on papermaking which was later published in the Transactions of this Society.[1] Doctor Franklin describes in his essay the laborious method used in Europe for forming large sheets of paper, and then relates the simple manner employed by the Chinese papermakers in solving the difficulty. The European method consisted of pasting small sheets together at the edges and burnishing the joints with a flint. In China a large mould was used, operated by two men and balanced by a counterpoise. The large sheets were couched upon the flat inclining sides of a heated oven or kiln, which gave one side of each sheet a remarkably smooth surface. Doctor Franklin states that by this method the Chinese workmen could make paper "four and an half ells long by one and an half ells wide."

As previously stated, it will not be possible to carry these de-

1 Description of the process to be observed in making large sheets of paper in the Chinese manner, with one smooth surface. In: Transactions of the American Philosophical Society . . . for promoting useful knowledge. Philadelphia, 1793. Volume 3, pages 8–10. 19 x 25 cm.

By the King and Queen,

A PROCLAMATION

For Publishing the several Marks or Stamps to be used for Velum, Parchment and Paper, pursuant to a late Act of Parliament for Charging certain Duties thereupon.

MARIE R.

God save King William and Queen Mary.

London, Printed by Charles Bill, and the Executrix of Thomas Newcomb, deceas'd; Printers to the King and Queens most Excellent Majesties. M DC XC IV.

NO. 66

A proclamation, dated 1694, relating to the various stamps to be used for vellum, parchment, and paper

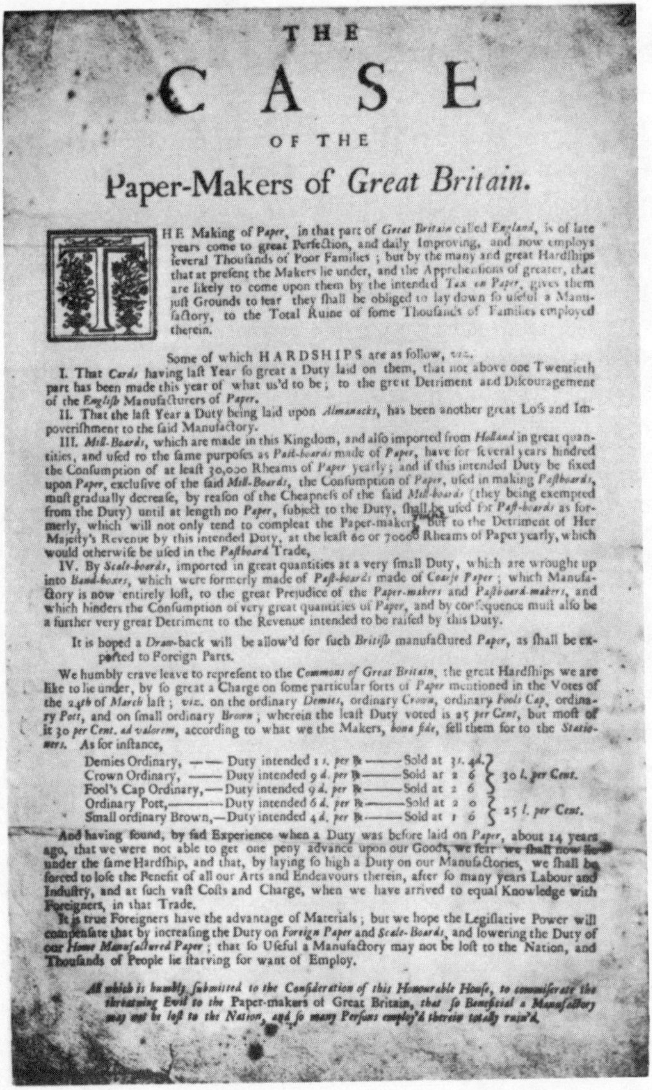

THE CASE OF THE Paper-Makers of *Great Britain.*

THE Making of *Paper*, in that part of *Great Britain* called *England*, is of late years come to great Perfection, and daily Improving, and now employs several Thousands of Poor Families; but by the many and great Hardships that at present the Makers lie under, and the Apprehensions of greater, that are likely to come upon them by the intended *Tax on Paper*, gives them just Grounds to fear they shall be obliged to lay down so useful a Manufactory, to the Total Ruine of some Thousands of Families employed therein.

Some of which HARDSHIPS are as follow, *viz.*

I. That *Cards* having last Year so great a Duty laid on them, that not above one Twentieth part has been made this year of what us'd to be; to the great Detriment and Discouragement of the *English* Manufacturers of *Paper*.

II. That the last Year a Duty being laid upon *Almanacks*, has been another great Loss and Impoverishment to the said Manufactory.

III. *Mill-Boards*, which are made in this Kingdom, and also imported from *Holland* in great quantities, and used to the same purposes as *Past-boards* made of *Paper*, have for several years hindred the Consumption of at least 30,000 Rheams of *Paper* yearly; and if this intended Duty be fixed upon *Paper*, exclusive of the said *Mill-Boards*, the Consumption of *Paper*, used in making *Pastboards*, must gradually decrease, by reason of the Cheapness of the said *Mill-boards* (they being exempted from the Duty) until at length no *Paper*, subject to the Duty, shall be used for *Past-boards* as formerly, which will not only tend to compleat the Paper-makers, but to the Detriment of Her Majesty's Revenue by this intended Duty, at the least 60 or 70000 Rheams of Paper yearly, which would otherwise be used in the *Pastboard* Trade,

IV. By *Scale-boards*, imported in great quantities at a very small Duty, which are wrought up into *Band-boxes*, which were formerly made of *Past-boards* made of *Coarse Paper*; which Manufactory is now entirely lost, to the great Prejudice of the *Paper-makers* and *Pastboard-makers*, and which hinders the Consumption of very great quantities of *Paper*, and by consequence must also be a further very great Detriment to the Revenue intended to be raised by this Duty.

It is hoped a *Draw*-back will be allow'd for such *British* manufactured *Paper*, as shall be exported to Foreign Parts.

We humbly crave leave to represent to the *Commons of Great Britain*, the great Hardships we are like to lie under, by so great a Charge on some particular sorts of *Paper* mentioned in the Votes of the 24*th* of *March* last; *viz.* on the ordinary *Demies*, ordinary *Crown*, ordinary *Fools Cap*, ordinary *Pott*, and on small ordinary *Brown*; wherein the least Duty voted is 25 *per Cent*, but most of it 30 *per Cent. ad valorem*, according to what we the Makers, *bona fide*, sell them for to the *Stationers*. As for instance,

Demies Ordinary, ——	Duty intended 1 *s. per* ℞ ——	Sold at 3 *s.* 4 *d.*	30 *l. per Cent.*
Crown Ordinary, ——	Duty intended 9 *d. per* ℞ ——	Sold at 2 6	
Fool's Cap Ordinary, —	Duty intended 9 *d. per* ℞ ——	Sold at 2 6	
Ordinary Pott, ——	Duty intended 6 *d. per* ℞ ——	Sold at 2 0	25 *l. per Cent.*
Small ordinary Brown, —	Duty intended 4 *d. per* ℞ ——	Sold at 1 6	

And having found, by sad Experience when a Duty was before laid on *Paper*, about 14 years ago, that we were not able to get one peny advance upon our Goods, we fear we shall now lie under the same Hardship, and that, by laying so high a Duty on our Manufactories, we shall be forced to lose the Benefit of all our Arts and Endeavours therein, after so many years Labour and Industry, and at such vast Costs and Charge, when we have arrived to equal Knowledge with Foreigners, in that Trade.

It is true Foreigners have the advantage of Materials; but we hope the Legislative Power will compensate that by increasing the Duty on *Foreign Paper* and *Scale-Boards*, and lowering the Duty of our *Home Manufactured Paper*; that so Useful a Manufactory may not be lost to the Nation, and Thousands of People lie starving for want of Employ.

All which is humbly submitted to the Consideration of this Honourable House, to commiserate the threatning Evil to the Paper-makers of Great Britain, *that so Beneficial a Manufactory may not be lost to the Nation, and so many Persons employ'd therein totally ruin'd.*

NO. 67

A broadside of the seventeenth century advocating a duty on paper

scriptions beyond the year 1800, as from this date books and pamphlets on the subjects of papermaking and watermarking become so numerous that it would require an entire volume to describe them. Perhaps the most important of modern works on watermarks is the four-volume dictionary compiled by Charles Moïse Briquet[1] (1839–1918) and originally published in Paris in 1907. The collector will not venture far without finding that the late Doctor Briquet made a valuable contribution to the history of watermarks. His dictionary is but the ripe fruit gathered from some twenty-three books on the subject compiled by him. By far the rarest of the modern works is the monumental work of H. T. Lichacev[2] which was published in St. Petersburg in 1899. In this work thousands of watermarks are reproduced in the three thick volumes and the folio atlas that comprise the set. For the most part, the literature of watermarks has been compiled in thin, paper covered pamphlets which will be found to be decidedly elusive when a collection is started. While it is not possible to state exactly the number of books and pamphlets which touch on the history of papermaking and watermarking, it would be safe to say that a complete bibliography would embrace several thousand items. There is no public library either in Europe or America that has even a representative assortment of works of this kind and their collection daily becomes a more difficult task.

1 Les filigranes; dictionnaire historique des marques du papier des leur apparition vers 1282 jusqu'en 1600; avec 39 figures dans le texte et 16,112 facsimilés de filigranes. Paris, 1907. 4 volumes.

2 Palaeographical importance of watermarks. St. Petersburg, 1899. 4 volumes and atlas. 22 x 32 cm.

PART II

The various stages of European papermaking, from the forming of the sheet to the finishing of the paper. From an engraving of 1762

PART II

THE APPLIANCES AND METHODS OF ORIENTAL AND OCCIDENTAL PAPERMAKERS DURING THE DEVELOPMENT OF THE CRAFT

TO trace the beginnings of papermaking we must go back almost two thousand years,—to the Chinese who were the inventors of paper. It is not known exactly when the craft was first practiced by them,—some historians state emphatically that it was in the year A.D. 105, while other antiquarians contend that the Chinese were forming sheets of paper as early as 123 B.C. Fortunately for the purpose of this book it will not be necessary to arrive at precise conclusions regarding this point. Papermaking was without doubt of Chinese origin, the art travelling west, first to Samarkand about the year 750, thence to Bagdad (793) and Damascus, reaching Morocco about 1100. Within fifty years after the craft was introduced into Morocco it extended into Europe, its inception being in Xativa, Spain; thence moving to France, where the first papermaking was done in Hérault in 1189; and on to Germany, being introduced either at Cologne in 1320 or at Nürnberg by Ulman Stromer in 1390. The art was probably practiced in Italy earlier than in Germany, for it is recorded that paper was made in Montefano as early as 1276, and later in Venice. In England paper was fabricated in the John Tate mill in 1494, but not in Holland until almost a hundred years later; and two hundred years after the first English mill was established the art of papermaking was introduced into the New World by William Rittenhouse.

The methods that were used in the making of paper during the early centuries of the craft are naturally vague after a lapse of almost two thousand years. No records indicate to us the manner in which the Chinese first made paper; but as with the ancient arts of the weaver and the potter, the method of hand fabrication in this craft must have changed but little from its inception to the present day. Almost two thousand years ago sheets of paper were made from disintegrated fibre and upon flat moulds, and in this fashion

NO. 68

The making of paper in the Southern Shan States. From a modern drawing executed by a native artist

paper is still formed, the only difference lying in the treatment of the fibre and in the construction of the moulds; in principle the process remains unchanged. The original papermakers of China probably poured the fibrous pulp into the box-like moulds, but it was not many centuries before the craftsmen practiced the more expeditious method of dipping the moulds directly into the liquid material. This change of method was one of the important steps in papermaking technique, as it enabled the artisans not only to form better sheets of paper, but to produce them in greater quantity. There is no way of arriving at the exact period when moulds were first dipped, but it probably did not take the ingenious Chinese a great while to discover the new and more efficient method. The original process of ladling the pulp into the moulds is not at present practiced anywhere in China, but in the Southern Shan States of Burma there is in use a method of moulding paper that would

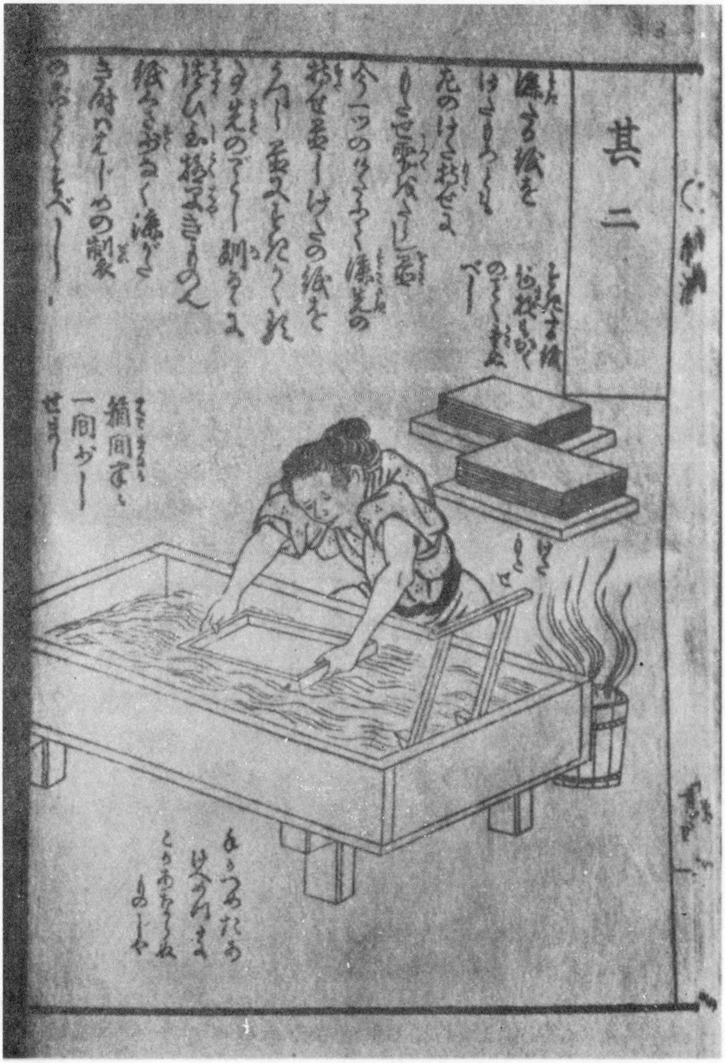

NO. 69

A page from the book by Kunihigashi Iibei, showing the mould being dipped into the vat, rather than the material being thrown into the mould correspond closely to the most ancient Chinese practice. This, however, could hardly be classed as a survival of the original Chinese system, as it is doubtful if paper has been made in this way in Shan above several hundred years. In the Southern Shan States the principal papermaking localities are Möng Kung, Lawksawk, and Kengtung, and their manner of moulding the sheets of paper is unlike that of any other locality. After the bark is prepared and macerated, a sufficient quantity for the making of one sheet is placed in a large receptacle half filled with water, and worked with a cross-forked pole rolled between the hands. This action separates the fibres and renders them suitable for making into paper. The actual sheets are then formed by throwing this pulpy mass into an oblong wooden frame-work of shallow construction, over the lower side of which is stretched a piece of native cloth or drill, forming a sort of loose sieve which allows the water to pass through freely, at the same time retaining the fibres. The mould containing the pulp is held in a trough, a little below water level, and upon bamboo supports. The pulp is quickly mixed with the water by the workers until it fills the cloth-bottomed mould to its boundaries. After the fibres have been well spread the mould is lifted from the trough, the thin layer of moist paper lying evenly upon the woven cloth. The worker then rests the filled mould upon the top edges of the trough to drain, the process requiring about ten minutes for each sheet. During the moulding process the workers throw out any unbeaten fibres or particles of dirt which would cause the paper to be uneven in texture or colour. After draining, the frame-work, with the wet sheet of paper still clinging to the fabric bottom, is placed to dry in the sun in an inclined position, supported by a pole of bamboo. When the sheet is partly dry, a worker rubs it with a piece of porcelain, a broken cup or bowl, which gives to one side a fairly smooth surface. After four to six hours exposure to the heat of the sun, the moisture has thor-

NO. 70
A Japanese paper mill of the eighteenth century

oughly evaporated and the sheet can easily be removed intact. The moulds used in Shan are about 24 by 60 inches in size, so that every sheet of paper formed on them is of the same dimensions. In the Möng Kung State, where the best paper of Shan is fabricated, the workers mix a little ground mica with the paper stock or pulp, which gives the surface a rather glossy white appearance, highly esteemed by the native paper users. The paper of the Southern Shan States is used for wrapping purposes, for exercise books, which are known as "parabaik," and for mats upon which to sleep, called "kalats." Their paper is sometimes made waterproof by rubbing the surface of each sheet several times with the oily substance of the "mai-tat-tao," (called in Burma, te-thee). It is my belief that the methods of making and finishing paper in use today in the Shan States are closely allied to the most ancient Chinese practices, though in no sense a survival of them.

There are no early Oriental books dealing with the actual fabrication of paper, so that it is not possible to determine from this source anything regarding papermaking in the East in ancient times. The earliest technical treatise upon Japanese papermaking was probably "Kamisuki Chohoki," the work by Kunihigashi Iibei, who made the first paper in his native province, Sekishu. The Oriental methods of forming sheets of paper were all identical in principle, but in the different countries, and even in various provinces of the same country, the equipment and manner of working differed to some degree. The methods of operation as shown in the Japanese book by Kunihigashi Iibei were practically the same as had been in use in Japan, as well as China, for hundreds of years, and show clearly that the moulds were dipped *into* the liquid pulp, rather than the pulp placed upon the moulds. According to the old book by Kunihigashi Iibei the material that was used for the making of paper was the "makaso" or "takaso," both species of the paper mulberry. The heavy paper that was

NO. 71

A Japanese papermaker dipping the mould into the vat. By the renowned Japanese artist Sadahidé, end of the eighteenth century

formed by men was termed "sugihara," while a lighter weight substance made entirely by the labour of women was called "hanshi."

Whatever the methods employed in the Orient at the inception of papermaking, the method used at the introduction of the craft

NO. 72

A Chinese paper mill. From, "Arts, métiers et cultures de la Chine, représentés dans une suite de gravures exécutées d'après les dessins originaux de Pékin, accompagnés des explications données par les missionnaires français et étrangers, pensionnés par Louis XIV, Louis XV, Louis XVI." Paris, 1814

into Spain during the twelfth century was to dip the mould into a vat filled with macerated liquid pulp, bringing it to the surface laden with the matted and felted fibre, which, after drying, was paper. In the four centuries which followed papermaking developed little. There was comparatively little use for paper in Europe before the invention of printing from movable types, but with the ushering in of this art the craft of the papermaker had its real impetus. It was during the fifteenth century that the fabrication of paper developed into a notable industry, and present-day interest in old papermaking dates from the infancy of printing rather than from the remoter periods.

NO. 73
The woodcut of 1568 by Jost Amman. The first delineation of a papermaker

The precise methods used by papermakers in the days of Gutenberg, Jenson, and Caxton are no more clear to us than are the exact processes employed by the early printers in their typemaking. The oldest print that represents an interior view of a paper mill is that of Jost Amman, from the year 1568, but this engraving throws no more light on the details of papermaking than does the print of the old typefoundry, in the same volume and by the same artist, show us how type was made at that time. It is only by piecing together material from many and varied sources that we are able to arrive at the *modus operandi* of the early craftsmen. The art of papermaking was jealously guarded by the old workers, and, there being little intercourse between the different mills, it was natural that the papermakers in each locality should have had their own methods of fabrication. Even today there is a secrecy among the makers of handmade paper in both the Orient and Occident, and the various mills show a pronounced variation in equipment and methods of operation.

In an examination of old Asiatic and European paper it may be readily perceived that the sheets were formed upon flat moulds; and from the impressions left in the paper by the material of which these moulds were made, it is easy to determine how the moulds were constructed. The moulds differed slightly in workmanship in the various countries and localities, and therefore we have inherited in documents, manuscripts, and books, innumerable varieties of paper, much beyond classification; yet it is clear that the fundamental principles of both mould construction and the actual making of the paper have undergone few changes since the original invention by the Chinese.

In all old paper mills, aside from the appliances that were used in the preparation of the pulp or paper stock (described in PART III), there were four essential utensils used,—the vat, the moulds, the felts, and the press. In every antique engraving depicting the papermakers' art these four necessary appliances will be seen. In the fifteenth century the vat for holding the fibrous liquid was simply a round or oblong wooden tub, about five feet in breadth, reaching in height to the worker's waist, and bound with stout hoops. The old vats were probably made from huge wine casks or tuns cut through the centre, as it is not likely that there were special shops where papermaking equipment was manufactured;

NO. 74

The vatman and coucher. From an engraving in *Theatrum Machinarum* by Georg Andreae Böckler, Nürnberg, 1662

doubtless the mills designed and constructed their own appliances to suit their individual needs. The earliest engraving representing papermaking shows no appurtenances to the vats, but during the seventeenth century several inventions were made which greatly facilitated the work. The most important of these was the introduction of the charcoal heater which was annexed to the back of the vat to warm the stock; by the more rapid evaporation which resulted the workers were enabled to make a greater number of

NO. 75

Papermaking in the seventeenth century. From *Visible World,* London, 1658

sheets in a day. In France this heater was known as a "pistolet," and it was the practice to place the vat against an outside wall so that the smoke from the heater or grate would not permeate the vat-house and soil the paper during the moulding process. Another improvement in the equipment was the "tripan," which was a platform or bridge extending from one side of the vat to the other. This bridge, as shown in the old engravings, supported the "horn" against which the mould with its newly-formed sheet was placed to drain. The bridge does not appear in the engraving of 1568 by Jost Amman, nor in the illustration of 1662 by Böckler, but inasmuch as it does appear in the small woodcut of 1658, it was evidently in use in Böckler's time in some parts of Europe. The pistolet and bridge are clearly delineated in the engraving in the work of de Lalande and also in the Basle woodcut (illustration No. 77), from about the same period.

During the process of moulding the sheets of paper it was essential that the fibrous material be kept from settling to the bottom of the dipping vat. Originally this was accomplished by the action of a pole in the hands of a workman, a means of agitation still used in some Oriental mills. A slight improvement was later made by placing on the end of the pole a wooden disc pierced with holes which increased the effectiveness of the stirring. Suggestions for more efficient work came slowly in the early mills, and each step, which now seems trivial, was no doubt welcomed as an ingenious invention by the old craftsmen. It was not until the eighteenth century that a mechanical agitator was devised; even this went through various changes before the paddle-wheel, called a "hog," was finally adopted and used almost universally.

The first method of supplying the vat was by carrying the macerated paper stock in buckets directly from the storage chest, a most tedious procedure; it was not until the seventeenth century that a supply by gravity was devised. I once visited a paper mill that had been equipped in the late eighteenth century, where the fibrous liquid had been conveyed to the vat from a large stone cistern by means of a long, winding wooden trough; the stock or pulp had been lifted and thrown into this trough by a huge wooden wheel fitted with copper buckets.

The vat filled to within a few inches of its brim with the diluted linen and cotton fibres which had been macerated under the stamping mills, things were in readiness for the actual making of paper. A workman, called a vatman, stood on a platform in front of the vat, holding a mould firmly by its narrowest sides; around this mould was a removable "deckle" or frame which acted like a shallow fence around its edges. (Moulds and deckles are described in detail in PART IV.) The vatman plunged the mould, at an almost perpendicular angle, into the liquid substance, and when the mould was well submerged he turned it face upwards and lifted

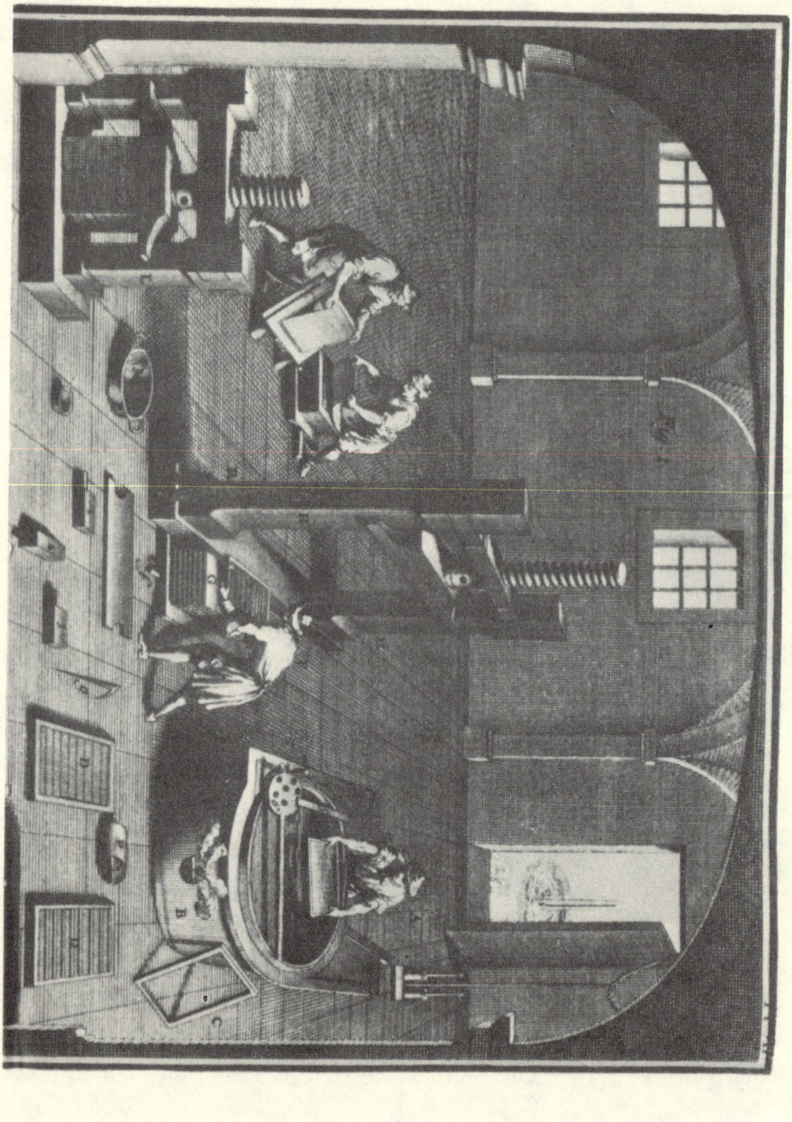

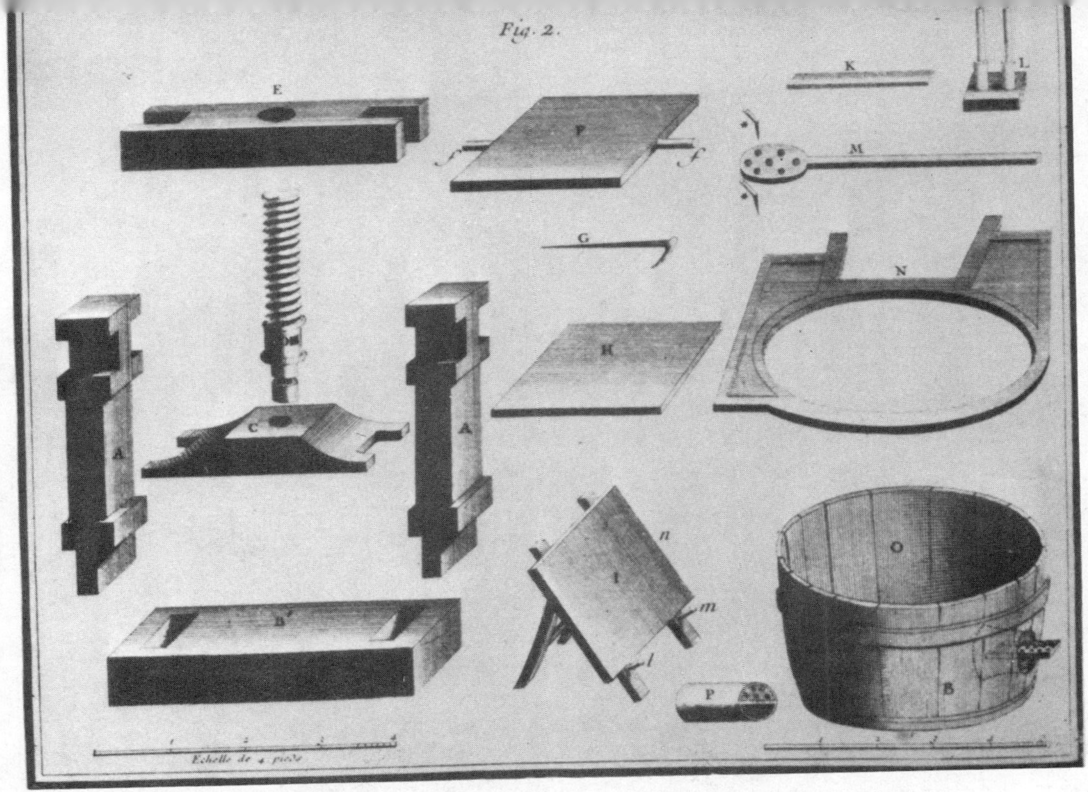

NO. 76

The moulding of paper and the various appliances as shown in the work on papermaking by Joseph Jérôme Le Français de Lalande

NO. 77

The complete process of forming sheets of paper showing the vatman, coucher, and layman at work. The original woodcut, from the end of the seventeenth century, is in Basle, Switzerland. The engraving has never been used in any book on papermaking

NO. 78
The rocking agitator was a great improvement over the stirring-pole. This means of preventing the papermaking material from settling in the vat came into use about the year 1800 and was used in most paper mills until the invention of the rotary agitator

it horizontally from the vat. By this action the mould was evenly covered with the fibrous material, and the stock not needed for the particular weight or thickness of paper being moulded, was allowed to run over the back edge of the mould into the vat, to be

NO. 79

A French paper mill of the eighteenth century. From the original painting in the Musée Lorrain

used again. The vatman then shook the newly-formed sheet on the surface of the mould, first from right to left, then from back to front. These motions crossed and matted the fibres, making the sheets of paper almost equally strong in all directions, and at the same time expelling considerable water from the sheet, the mould being so constructed that it retained the fibres but not the water. The vatman then removed the wooden deckle, which left the moist sheet of paper, cut sharply along its four edges, upon the mould. The next operation of the vatman was to pass the mould along the bridge to the second artisan, known as the "coucher," and proceed to form another sheet with a second mould, using the same deckle. The coucher placed the mould, with its thin moist sheet, against the inclined drainage-horn, which allowed the surplus water to drain back into the vat. When the wet sheet of paper had solidified upon the mould to the proper extent,—a stage determined by a distinctive lustre over its surface—the coucher turned the mould completely over and deposited the wet sheet of paper upon a piece of felting. The couching was done with a quick, deft motion, rocking the mould from one long edge to the other, leaving the spongy sheet, flat and unwrinkled, upon the rough woolen cloth. These processes were repeated over and over, the vatman and coucher working together, until a pile of 144 sheets of paper had been formed and couched, each sheet of wet paper separated from the next by a piece of felt somewhat larger than the size of the paper. The pile of six quires was termed a "post" and the next operation was to place the post in a press to expel the excess water.

In the earliest form of Chinese papermaking the sheets were dried directly upon the cloth moulds upon which they had been formed, so that couching, or removing the paper from the mould while wet, was not required. After the introduction of the rigid mould, when dipping was used, it became the practice to couch each sheet as it was made upon matted wool or felting, but for a long period the wet sheets of paper were couched one upon the other without anything between them. This method is in use at the present time in some parts of Asia.

At the introduction of papermaking into Europe felt of some kind was in all probability used as a material for the couching of paper. It is not likely that the old craftsmen employed woven wool cloth such as the makers of handmade paper use at present, but a more compact, matted substance of hair or wool. The woven material that is used by modern papermakers is called felt, but the name is a misnomer as felt proper is a compressed mass and does not consist of warp and weft. Years ago in Catholic countries the

NO. 80

The great cumbersome press constituted one of the most expensive appliances required by the old papermakers. This press was used in a paper mill in Haynsburg, Germany, from 1700 until 1909 when it was moved to the Deutsches Museum, München, where it is now on exhibition

felt-makers celebrated the 23d of November, St. Clement's Day, with a great festival. It is stated that this Saint, when on a pilgrimage, placed carded wool between his feet and the soles of his sandals, and at the end of his journey found that the wool had been converted into felt. However, the invention of felt by St. Clement is to be questioned, as this material was in use long before his time. The Greeks manufactured true felt as early as 900 B.C., which was before the arts of weaving and papermaking had been introduced to civilization. Papyrus was, it is true, made before this period, but this is a laminated substance and was not made from macerated and disintegrated fibre as is true paper. Both the Greeks and

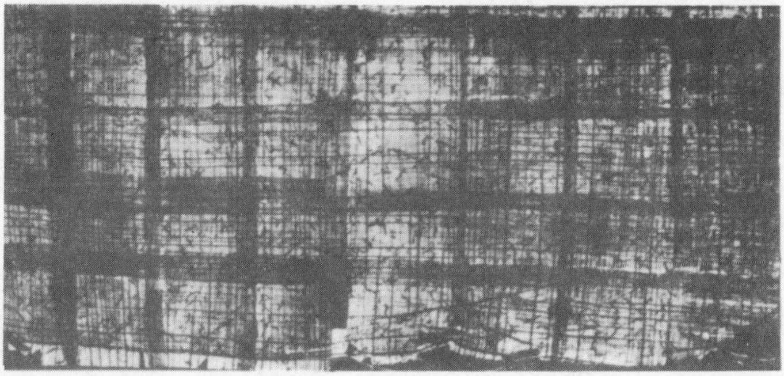

NO. 81

Papyrus was not true paper composed of disintegrated fibre, but a laminated substance

Romans made use of matted felt for caps, blankets, and the lining of helmets. Since the material was thicker than common cloth it presented a more effectual obstacle to missile weapons; hence, when the soldiers under Julius Caesar were annoyed by Pompey's archers, they made shirts of heavy felt and wore them for defence. The Scythians, who from the most remote time roamed the lands of Northern Asia, as did the Tartars, their successors, used felt for both clothing and material for the making of tents. It is therefore quite evident that had the early papermakers, even the original Chinese, desired felt for the couching of their paper it could have been readily procured.

It is not known when felt came into use as a material for receiving the wet sheets of paper as they were couched from the moulds, nor is it known when woven wool cloth was substituted for the matted felting for the couching of paper, but it is probable that some sort of woven material was in use for the purpose as early as the thirteenth century. No special wool cloths were fabricated for papermaking until the eighteenth century, although there was an incorporated company of felt-makers in England as early as 1604. These artisans made felt for hats, but it is not recorded that they attempted any woven or matted material especially for the papermakers. According to the diarist, Samuel Pepys, a lecture on the subject of felt and felt-making was given by Dr. Hooke before the Royal Society of London in 1666. The material used for the making of felt was not only wool, but the hair of the seal, rabbit, monkey, goat, and camel.

The best description of woven cloth such as was used in the eighteenth century by the makers of paper is furnished by Nicolas Desmarets in *Traité de l'Art de Fabriquer le Papier* published in Paris in 1788:

"The felts have two surfaces furnished with different naps. That side which has the longest nap is applied to the couched sheets, and on the side with the shortest the fresh leaves are laid. If this arrangement of the felting was changed, and the sheets laid upon the surface with a long nap, not only would they not apply themselves accurately to the felt, but the long stiff hairs would pierce the paper or cause depressions, which would injure the texture. On the contrary, the leaves fit themselves evenly to the side with a short nap, which absorbs the surplus water, and gives a sufficient consistency for the coucher. It is also from this side that the layman detaches the sheets after the post has passed through the press so that the different character of these surfaces is an assistance to the layman as well as to the coucher. The material of which the felting is made requires great care on the part of the papermakers, and much attention and knowledge of the art of papermaking on the part of the manufacturer who prepares them. They should be firm enough to spread evenly upon the leaves without wrinkling or needing to be displaced. Also, they should be supple enough to adapt themselves to the work of the coucher, who applies the mould successively from one edge of the felt to the other upon every intermediate point. As the felts have to resist the reiterated efforts of the coucher and the press, it is necessary that the warp of these cloths should be very strong and therefore made of combed and well-twisted wool. And as they should be clean and quickly absorb and give up again a certain amount of water, their weft should be of carded wool quite loosely spun and woven in about the same manner as light cloth. It results from this that the weft abundantly fills the stuff and covers the warp in such a way that the texture is not marked upon the paper, which would injure its grain by the irregular impression of an uncovered warp and weft, as is often observed in establishments where felts not woven upon these principles are used, and which does not denote a great deal of talent on the part of the papermaker. If the material were too closely woven, like ordinary cloths, or even the finest kinds, it would not absorb the water enough to enable the sheets to adhere, and assume a certain consistency. It is for this reason that Carcassonne cloths are very well adapted to this purpose, and that those of Louviers, of which the texture is very close, would not take the leaves of paper couched upon them in experiments several times repeated, because the water could not sufficiently penetrate them. It is very essential that the warp of the stuff, intended to be made into felts, should be strong and tough, so that they may be of good service, and wear well."

For pressing the water from the piles of paper and felts a wooden screw press such as is portrayed by Jost Amman in 1568 and Georg Andreae Böckler in 1662 was employed in the old mills. As these presses had to yield an immense pressure they were massive and cumbersome, and constituted one of the most expensive appliances of the old paper mills. After the post had been placed under the platen of the press, all of the workers in the various parts of the mill were summoned together by the ringing of a bell so that they might exert their strength in turning down the screw of the press by means of a long wooden lever. As the platen of the press closed

NO. 82

The vatman. From an engraving in *Iets Voor Allen*. Amsterdam, 1717

downward upon the felts and newly-formed paper, an abundance of water was forced from the mass, the post being reduced from about two feet in thickness to barely six inches. In later years a device called a "Samson" or "sampson" was brought into use in the larger mills. This device acted as a windlass, and with the power of six men the pressing was more efficient than had been previously possible, giving the paper a more compact texture. To insure a uniform pressure on the pile of paper and felting the workmen placed thin bevelled boards around the top of the post, for without this precaution the centre would have received more pressure than the edges. The results of these apparently simple innovations are reflected in the old papers, and without a knowledge of the different methods of working it is not possible to determine the age or origin of antique papers.

NO. 83
A paper mill interior of the late eighteenth century, showing the "Samson" used in screwing down the press

After pressing, the weight of the paper and felts, from the loss of the water, had been reduced tenfold, and the fibres of the sheets of paper had been so matted and felted together by the great pressure that the sheets were lifted from the woolen cloths without tearing, each sheet being a homogeneous piece of paper, although still not entirely dry. The third workman, in the process of old papermaking, was known as the "layman," and it was his duty to free each sheet of paper from the interleaving felts and to place the sheets in a neat and even pile, one upon another, on an inclined stool or bench. The felts were returned to the coucher so that they might be used in making the following post; these pure wool cloths were expensive and the old mills had but one set for each size of paper that they made. By long hours and diligent work it was possible for the three workmen to make twenty posts, or about five and one half reams of paper, a day. Had but a single mould been used, one man acting as both vatman and coucher, scarcely one-fourth of this amount would have resulted.

After the sheets had been placed one upon another by the layman, the pile was again subjected to pressure, but with nothing between the sheets. This process, although of light pressure, brought out a little more water, which formed in drops upon the edges of the sheets. After this second slight pressing the sheets were separated, built in a pile again, but in different rotation, and subjected to a little heavier pressure than could have been given before, owing to the tendency of the sheets to stick together. This exchanging was repeated until the paper acquired the desired smoothness of finish. The roughness of much of the paper previous to the sixteenth century would indicate that before that time this exchanging or parting was not in practice in some localities, the paper evidently having received no finishing after removal from the woolen felts. In an examination of old paper, especially Italian and Dutch, it will be noticed that many of the sheets retain the impressions left by the material upon which they were couched. The early American papermakers did not devote much time or labour to the exchanging, although this procedure was, and still is, the secret of giving to handmade paper a beautiful finish and texture.

The next process, after the exchanging, was the drying of the paper, which was one of the most particular operations of the mill.

NO. 84

The vatman, the coucher, and the layman at work. The engraving also shows the press and vat. From the work of 1774 by Nicolas Desmarets

NO. 85

The drying-loft of a paper mill. The paper is being hung upon ropes by the use of the "T," shown in Figure 5. This method of hanging paper to dry is still employed in handmade paper mills. From the engraving in the work of 1774 by Nicolas Desmarets

The paper was taken from the pile, after the last pressing, in "spurs" of four or five sheets, as the paper in this moist state adhered together. Had the sheets been dried separately they would have wrinkled, but when taken in spurs the paper dried smoothly and without excessive curling. The drying-lofts of the old paper mills were situated in the top stories of the buildings where the air was pure and free from particles of dirt and soot. The walls of these lofts were fitted with sliding shutters which could be shifted to permit or exclude the air. The spurs of paper were hung over ropes that had been woven from cow- or horsehair, coated with beeswax. These heavy ropes supported by horizontal wooden frames called "tribbles" were stretched throughout the loft, and the entrance of the air arranged so that it came in contact with the full breadth of the sheets and not against the edges of the paper. Ropes are still used in European mills for the drying of handmade paper, but the paper mills of America in the days of loft drying employed wooden poles for this purpose.

In the fifteenth and sixteenth century mills the drying was carried on in the most primitive manner, and it is not unusual in paper from this period to find stains through the middle of the sheets caused by the hair ropes upon which they had been hung to dry. During the development of papermaking in America some of the mills transported their paper to the printers before it had dried thoroughly, as the makers were not able to supply the demand with sufficient rapidity. In many folio volumes, particularly from the seventeenth century, the pages are bulged in the centres, causing the leaves to lie in ridges, somewhat in the shape of a diamond. While this defect is more noticeable in folios it may be found in smaller books also, and was caused by the paper not having been exchanged or pressed sufficiently; during the drying the weight of the paper stretched the sheets, causing them to bulge in the centres. The Oriental papers were dried flat against boards or

Fig. 1.

Partie des Etendoirs.

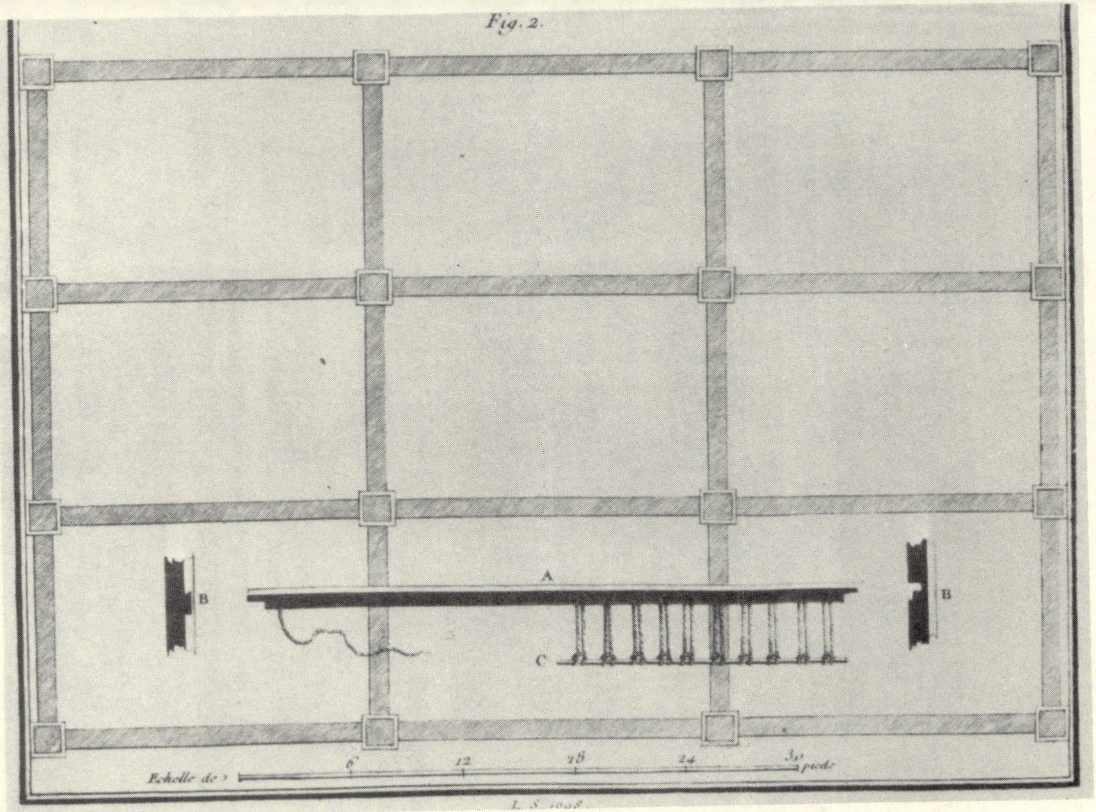

NO. 86

The drying-loft and its construction as pictured in the comprehensive work on papermaking by de Lalande

smooth masonry walls in the sun, or against the sloping sides of plaster ovens made especially for the purpose. In drying paper in this fashion the sheets were held rigid and flat, so that the old Oriental papers are usually smooth and without ridges.

NO. 87

The vatman, coucher, and layman. From a French print of the eighteenth century

The drying completed, the old papermakers dipped their paper into an animal size that had been made from the parings of hides, which they procured from the parchment-makers. It was necessary to size the paper so that it would be impervious to ink, but sizing was more needed in writing than in printing papers. Many books of the fifteenth century were printed upon paper that had not been sized, this extra treatment not being essential for a type impression. The sizing was accomplished by a worker holding a number of sheets by the aid of two wooden sticks, and dipping the paper into the warm glutinous liquid. The sheets were then pressed to extract the superfluous gelatine. This crude method of sizing the paper was extremely wasteful as many sheets were torn and bruised beyond use. The sizing room of the early paper mills, was, for this reason, known as the "slaughter-house." One of the earliest Oriental methods of sizing paper consisted of covering the surface of the sheets with a thin coating of gypsum. The next improvement was to render the body of the paper, as well as the surface, impermeable to ink by the use of lichen, starch, or rice flour. The Chinese used starch as a size for paper as early as A.D. 768 and its use continued until the fourteenth century when animal glue was substituted.

The sizing completed, the paper was again dried, and taken to a room which was called the "saul." In England the finishing room of a paper mill has been known by this name for several hundred years. The term may have come originally from the German word "saal," and it is reasonable to surmise that the phrase was first used in England by John Spilman, a German, who established a mill in Kent in 1588. The saul was that part of the mill where the paper was surfaced and packed for transportation. The earliest method of producing a smooth surface on the paper was to burnish each sheet by hand by the use of an agate or other glossy stone. The Orientals gave their paper a glass-like surface by this procedure. This burnishing was done on one side of the sheet only, but as the reverse side was laid against a smooth slab of marble or wood it also took on a certain amount of finish. Even when the Asiatic scribes made use of European paper they subjected it to a polishing with an agate or soap stone burnisher which gave the paper an Oriental appearance; this action also closed the pores of the sheets and rendered them more suited for writing upon with ink. A mechanical method of imparting a surface to paper was brought into use in Europe in the early seventeenth century. This

Papeterie
Fig. 1.
A
B
C
D
E
F
G
H
I
K
L
d
d

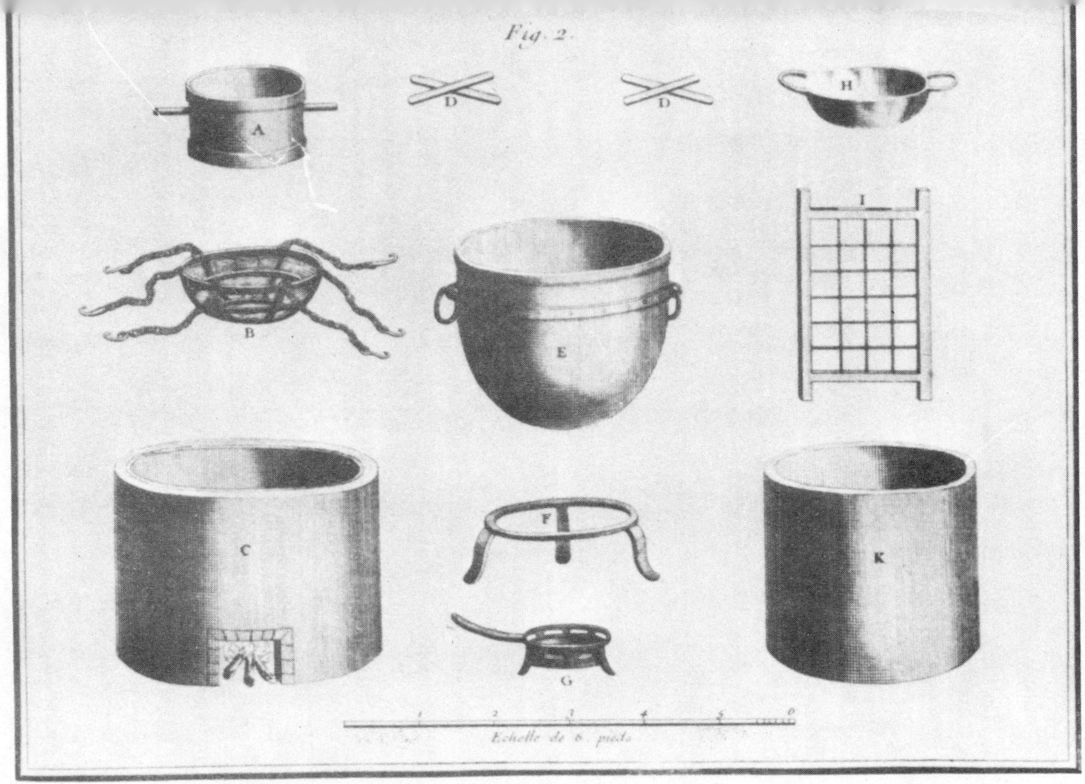

NO. 88

The sizing of paper and the appliances used in the process. From the eighteenth century work by de Lalande consisted of a pressing hammer of huge construction, the implement being operated by water power. The idea for the apparatus was borrowed by the papermakers from the bookbinders, as these craftsmen had employed this mechanism for polishing the backs

NO. 89

The interior of a German paper mill, dated 1601

of books many years before the papermakers adopted it. This machine gave the paper a more uniform surface than had been possible by the hand rubbing method. The results achieved by the use of the glazing-hammer were soon spread to the different mills by the travelling workers, and it was not long until this machine was used universally. The hand-burnishers looked upon the introduction of the hammer none too kindly, for when it came into use they had to seek other employment. In Germany the introduction of the glazing-hammer caused a division among the makers of paper. The mills that continued the use of the old hand-burnishers would not allow, under severe punishment, any artisan to work for them who had been employed in a mill where the new mechanical device was used. The controversy became so heated that the Government threatened to confiscate the tools used in hand-burnishing, and recommended that the mills adopt the new pressing- or glazing-hammers. In spite of this ruling, however, the mill owners continued to polish or finish their paper in the manner they desired. Polishing by hand always gave the paper a streaked

NO. 90

An eighteenth century engraving showing the various processes of papermaking,—rag sorting, fermenting, stamping, moulding, sizing, drying, and packing

Papeterie.
Fig. 1.

NO. 91

The inspection of finished paper and the glazing-hammer as pictured in *Art de faire le papier* by de Lalande

and uneven appearance owing to the narrowness of the glazing-stone, while the huge hammers imparted an almost uniform surface over the entire sheet. In studying old Oriental and Occidental papers these slight variations are traceable and afford somewhat of

NO. 92
Papermaker, 1786

a clue as to the origin of the paper. About 1720 the pressing-hammers gave way to the wooden glazing-rolls for giving the paper a smooth finish. This was a Dutch invention and consisted of two large rollers, cut from the solid trunk of a tree, and between which the paper was pressed. The finish which this machine imparted was as superior to that of the pressing-hammer as that had been to the original hand-burnishing by stone.

After the paper was finished it was packed for transportation by

NO. 93
Papermaking. From *The Sister Arts,* Lewis, 1809

NO. 94
The papermakers. An engraving dated 1817

wrapping in heavy coarse paper of a brown or grey colour; this wrapping paper was made from the dregs of the vat that could not have been used for the moulding of book or writing paper. While there are no paper-package labels from the fifteenth century ex-

NO. 95
Papermaking by hand. From a Dutch Lithograph of about 1800

tant, there have been a good many preserved from later centuries, and in another part of this book a representative collection is reproduced.

The methods at present used in making handmade paper differ but little from those described. Several improvements in the construction of the vats have been made, the chief of which was the "knotter," which eliminated foreign substances and knotted fibres from the paper stock before it ran into the vat. This appliance, an American invention, came into general use about 1819; before its introduction it was necessary to scrape out from the sheets the knots made by twisted fibres or by bits of broken buttons, pieces of stone, wood, metal, or earthy matter. These could not be removed without considerable injury to the paper, often causing holes in it. The knotter or strainer was greatly improved in 1832 by a Kentish papermaker, and it is this form of the appliance that is still in use in handmade-paper mills. Another improvement was

NO. 96

Papermaking. From *The Book of English Trades*, London, 1825

the use of a steam pipe for the heating of the stock in the vat in place of the old pistolet, while the primitive stirring-pole was superseded by an agitator which revolved in the bottom of the vat, keeping the liquid material in constant motion. The heavy wooden screw press may still be seen in several of the older handmade-paper mills in Europe, but the hydrostatic-press is now used by all well equipped vat mills. This was invented in 1790, but was not adopted until long after that date, although Joseph Bramah, the originator, recommended it for use in paper mills.

The method of drying handmade paper remains today very much as it was in the fifteenth century, but the old sizing tub has been replaced by a machine composed of a trough and rollers, each sheet of the paper separately being led through the warm size in the trough upon an endless felt. A century or more ago the heavy wooden rolls of the Dutch glazing machine gave way to iron, but the principle of the apparatus remained about the same as in the early eighteenth century original. Every appliance of a modern handmade-paper mill can be traced to its Oriental origin, the steps upward being separated by centuries. To be able to determine the locality and period of old papers it is necessary to make a minute study of the ancient papermaking appliances, as the succession of innovations, occurring in different countries at various periods, caused changes in the character of the papers throughout the ages. To the casual observer these changes are insignificant, but to the student of papermaking they are clues which, followed, lead to secrets of the craft he loves.

PART III

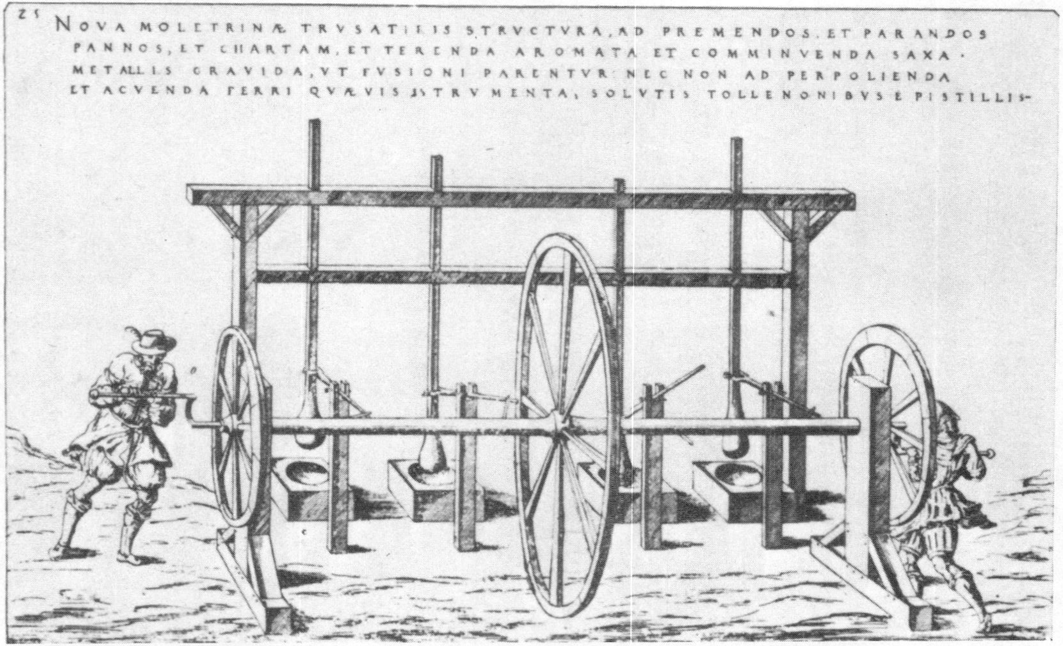

An appliance for the maceration of papermaking material as depicted in *Théâtre des Instruments mathématiques et mécaniques de Jacques Besson, Dauphinois, docte mathématicien.* Lyon, MDLXXIX

PART III

DEVELOPMENTS IN THE MACERATION OF PAPERMAKING MATERIALS, FROM THE ANCIENT CHINESE METHODS TO THE INVENTION OF THE "HOLLANDER"

IN the preceding chapter it has been explained how in both the Orient and Occident the paper was actually formed into sheets. It will now be interesting to trace the development of papermaking materials, but more especially their maceration; for all substances, no matter of what origin or nature, have first to be reduced to fibre before they can be suspended in water and formed into sheets of paper.

The earliest papermakers, the Chinese, may at first have used silk, both raw and woven, as material for their paper. Certainly it is significant that the Chinese character for paper (紙) is composed of two radicals, the first being radical 120, of six strokes (Ssû), meaning raw silk, and the second, radical 83, with four strokes (Shih), denoting sir or family. The original Oriental mode of beating or macerating the substance to a pulp was by placing the mass in stone mortars and pounding the fibres to a paste by hand with wooden mallets. About A.D. 105 Ts'ai Lun, whose mortar was still preserved as a curiosity in the time of the Thang Dynasty (A.D. 618–907), made the important discovery that vegetable fibres such as the bark of trees and the filaments of plants and grasses, could be substituted for silk, or animal, fibres. The definition of the word paper, given in the Shuo-wên, which was compiled about this time, would indicate that to the compiler the word applied not only to the pieces of woven silk fabric in use at the time by the Chinese for their calligraphy, but that it meant also a sort of paper or paper-like substance. The usual date given for the invention of papermaking in China, A.D. 105, would suggest that Ts'ai Lun's substitution of vegetable fibres for animal fibres marked the real beginning of the industry in China.

According to the researches of Doctors Wiesner and Karabacek, former Professors in the University of Vienna, cotton fibre in its

NO. 97

The maceration of papermaking material in the time of Ts'ai Lun

raw state was never used by the ancients as a papermaking material. The papers, dating from A.D. 709 to 909 which these gentlemen examined, were made by the Arabs from material composed almost entirely of linen. The Arabs learned the craft of papermaking in the year A.D. 751 from Chinese prisoners in Samarkand. In their own country the Chinese at this date used the bark of mulberry trees for their paper, but this material not being available in Samarkand, the Arabs employed linen as a substitute. To the Arabs of Samarkand, therefore, must be given the credit for the fabrication of the first linen paper. Cotton, either raw or woven, was apparently never used by the early Oriental papermakers.

The Arabs probably disintegrated the linen rags by placing them in heaps, saturating them with water and allowing a fermentation to take place, and then boiling the mass in wood ashes. The boiled rags were next placed in cloth bags and suspended in a running stream the action of which removed a great part of the alkaline residue and much of the dirt. The original method of beating the cleansed rags to a pulp was doubtless an imitation of the Chinese manner, but later in the development of papermaking by

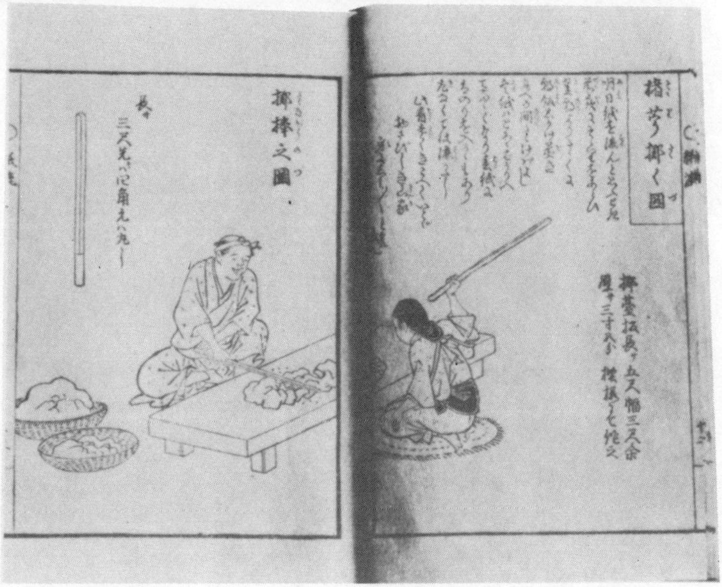

NO. 98

Beating papermaking material in Japan. From the rare book by Kunihigashi Iibei

the Arabs a trip-hammer was put to use. The workers treading upon the end of the horizontal tilt-bar of this implement caused the hammer to fall heavily upon the substance to be beaten. This required far less labour than the Chinese process, yet for hundreds of years after the improved method of maceration had been invented by the Arabs and adopted in various parts of Asia the hand-beating of the bark for papermaking was employed in both China and Japan. In Japan, as late as the time of Kunihigashi Iibei (eighteenth century), hand-beating was used to reduce the mulberry bark (makoso) to a pulp; the cleansed bark, laid on a heavy wooden plank about three by five feet in size, was pounded with square beating sticks, three feet in length, with handles rounded

NO. 99

In India there are several handmade-paper mills where the method of beating the papermaking materials is the same as that used in Samarkand in ancient times

to fit the hands of the workers. The Japanese paper described by Iibei was made in the province of Sekishu and usually in the winter, as that fabricated in the spring was not considered to be so fine.

In the paper village of Tonkin in French Indo-China the workers in paper are using at the present time the old mortar, with pestle wielded by hand, the ancient Chinese method. Their raw material is the bark of the mulberry tree, which after soaking and boiling in lime can readily be beaten in this way. In India at the present time there are several handmade-paper mills in operation where the ancient Samarkand method of beating is in use. The native artisans use as their raw materials jute, discarded gunny sacks, and old papers. These substances are cut to bits, placed in baskets and washed thoroughly in running streams until the fibres

NO. 100

Maceration of papermaking material in China. From, *Arts, métiers et cultures de la Chine,* Paris, 1814

have separated. The stuff is then put into stone vats where it is suffered to soak in lime for a period of several days. After this process is completed it is thrown into rough stone mortars sunk in the ground, where it is beaten by means of heavy pestles, like huge hammers, balanced on pivots. Men and boys stand on the long horizontal wooden beams, or handles of the hammers, and by their weight raise the pestles, letting them fall upon the jute or other material, over and over again, until the substance is beaten to a pulp suitable for forming into sheets of paper. The process is almost identical with that of Samarkand in use a thousand years ago.

At the introduction of papermaking into Europe the methods of the East were no doubt used for the maceration of the pulp. It did not, however, take the Europeans long to devise a better method of beating, for about 1151 there was invented at Xativa, an ancient city of Valencia, Spain, a stamping-mill operated by water power. At that period linen and cotton were used as material for making paper. For the preliminary treatment the well-worn cloth was wetted and pressed into balls, which were left in piles for six weeks or two months, being kept moist the while. This caused a fermentation, manifested by the increased temperature of the mass, the rags assuming a peculiar colour, not unlike the dregs of wine. This method of preparation was wasteful in the extreme as fully one-third of the material was rotted beyond use. The portion suited for paper was, however, tender and easily beaten, so that the action of the newly invented stamping-mills had immediate effect upon the rotted and disintegrated cloth.

During the initial centuries of papermaking in Europe the demand for the commodity was slight, but after the advent of printing from movable types in the middle fifteenth century the demand became greater, and by the close of the century the art of papermaking on the continent had developed beyond belief and had spread to many places where the craft had not hitherto been practiced. During this period linen was used in wastefully large quantities, as there was considerable loss of the material during the fermenting process and also in the actual stamping of the cloth into fibre. In the fifteenth century in Europe there was not of course the great variety of cloth that there is at present, so that the sorting of the rags was not a difficult problem as it is today; the linen and cotton fabrics of that period were naturally all woven by hand, and being free from chemicals and bleaching agents were ideal materials from which to fabricate paper. Linen of the whitest kind was desired, but cotton was not rejected. When materials of different weights and substances were used, they were, in the best mills, pounded or stamped separately and mixed after beating, for without this precaution much of the fibre would have been lost, the finer particles escaping with the water during the lengthy process of reducing the tougher and coarser fabrics.

In the mills of less importance where the rags were fewer, they were left to rot for longer periods, as the heaps, being smaller, did not heat so rapidly. The length of time for this process also depended upon the quality of the rags; the finest linen did not rot so quickly as the coarse, and the linen that had been worn as clothing rotted with more difficulty than new material, because the internal humidity that disposed the fibres to fermentation was more considerable in new or coarse than in worn or fine linen. When fungi grew on the heaps or piles of rags it was considered that the process was well under way. In some mills lime was thrown upon the rags to assist in the disintegration, but this was detrimental to the material, and paper made from material so treated was never so strong or lasting as that which had fermented naturally. In the eighteenth century the use of lime was absolutely prohibited in France. After the fermentation process the rags were cut and placed in stone or heavy wooden troughs where they were washed by hand, for during the rotting the rags took on a yellow colour which had to be eliminated. This was accomplished only with difficulty; in fact, most of the paper from the early mills shows a yellowish, but pleasing, tint, the direct result of the fermenting process.

After the cloth had been reduced to a broken and tender state by the rotting, and had been cleansed by washing, it was ready to be triturated into pulp for forming the actual sheets of paper. The stamping-mills that were originated in Spain consisted of rows of great wooden hammers, at first set at a slant, but later perpendicularly, which were made to rise and fall by means of a series of cams on a stout axis. These hammers or pestles operated up and down in troughs into which the rags were thrown; the troughs or "vat-

NO. 101

The original stampers were set at a slant, but later they were placed perpendicularly

holes," as they were termed, were fashioned from great blocks of stone or from logs of the oak tree, the wooden cavities being lined with iron or lead. These early European mills were turned by water power, which differentiated them from mills of the Orient where the beating had been accomplished by the power of man. While European workmen received from the Orientals the principles of papermaking, it was not long before they had put to use their own methods in the whole operation,—the rigid, metal-covered mould was devised, a better and less laborious process of beating was adopted, and the streams and rivers were put to work to turn their paper mills. Yet after these innovations were brought into use the craft experienced but few changes over a period of several hundred years; in the fifteenth and sixteenth centuries Eu-

NO. 102

Modifications of the stamping-mill were in use in Europe from the twelfth to the eighteenth century. This is the first presentation of a mill of this type, from the book on machinery by Vittorio Zonca, Padua, 1607

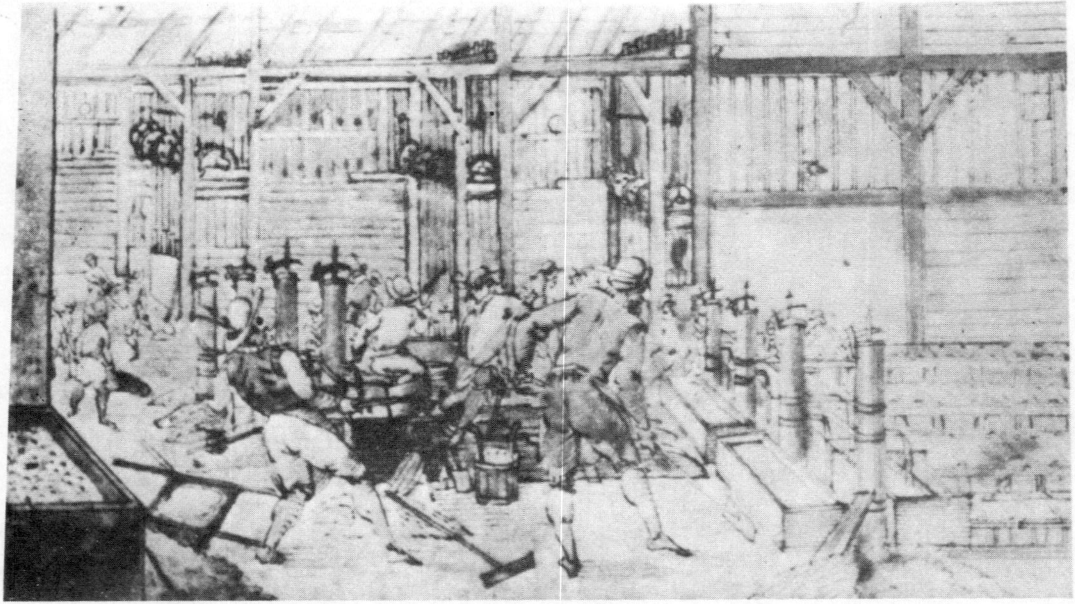

NO. 103

The preparation of papermaking material in the early seventeenth century. From the original drawing in colour, dated 1606, by an unknown German artist

ropean paper was made very much as it had been fabricated in the twelfth.

Toward the seventeenth century the mills did show slight changes, but the principle of the *modus operandi* remained unal-

NO. 104

A model of an old stamping-mill. The original is in the Town Hall Museum, Augsburg

tered. At this period the stampers were constructed on a larger scale and were a little more developed in their capacity, the stamps being divided into three classes. The first set of stampers or pestles were shod with rough iron teeth, or spikes, which frayed the cloth while a stream of pure water ran into each stone or wooden trough and cleansed the rags, the discarded dirty water flowing off through holes in the sides of the receptacles, over which was placed woven horsehair screening to prevent the escape of the disintegrated fibres. This procedure replaced the more laborious method of hand washing used at the inception of papermaking in Europe. When the rags were partially cleansed and somewhat mangled, they were bailed out of the first set of troughs and placed under the next set of stamps, which were less heavily shod. In the second operation the stream of pure water was continued to complete the washing process. The fresh water was supplied through a wooden leader to each trough or vat-hole, either by gravity from

NO. 105
A section of an old stamping-mill with a stone vat

a mountain spring or from the stream that furnished the locomotion for the stampers. The water was usually filtered through a series of large wooden reservoirs, each having an opening covered with iron bars placed closely together which eliminated the gravel and the larger particles of dirt and vegetable matter from the water. The final treatment of the partly beaten rags, or half-stuff, was given by the third and last set of stampers, which were usually made of plain wood, without iron or steel facing. At this stage the running water was excluded, for by this time the rags had become so disintegrated that much of the fibre would have been lost through the strainers, or washing screens.

It is recorded that Ulman Stromer, who established a paper mill in Nürnberg about 1390, "employed two rollers (axes) which set eighteen stampers in motion." In the Italian book on machinery by Vittorio Zonca, printed in 1607, the writer depicts for the first

NO. 106

The water used in the old stamping-mills was filtered through a series of wooden boxes, the apertures of which were covered with metal wire or horse-hair screens

time a set of these stampers. In his engraving there are eight stampers to one receptacle, and in the volume by Georg Andreae Böckler, dated 1662, the same number is delineated in the plate showing this machine. The German picture was apparently inspired by the earlier Italian drawing, the two being almost identical. In the other plate, however, in the Böckler book, three stampers are shown to each trough. As late as 1718, according to the work by Leonhardt Christoph Sturm, there were five sets of stampers to each stone or wooden trough and four troughs at each side of one axis, the axis and its cams actuating forty separate

NO. 107

A stamping-mill of the seventeenth century. From *Theatrum Machinarum* by Georg Andreae Böckler, Nürnberg, 1662

stamps, and one water wheel turning the entire apparatus. The stamping-mills were slowly developed from the crude and cumbersome machine such as Zonca pictured, until a good German mill had as many as twenty-five troughs with four stampers to each trough, while French machines had from three to five stampers to every receptacle. These stamping-mills were used for the maceration of rags by the early American papermakers, but their use was discontinued in this country a number of years before they were abandoned in Europe. The Willcox mill was not established until 1727, but it is recorded that their first method of reducing rags to pulp was by the use of a stamping-mill. This was the third paper mill set up in Pennsylvania. It is not known when the

NO. 108

A stamping-mill such as was used in America in the early years of papermaking. From the engraving in the work by M. Desmarets

stampers were discarded in America, but it is thought that they remained in action until about the year 1780. With the use of these old machines the fibres of the cloth were slowly drawn out, almost to full length, which is one of the reasons the papers of the stamping-mill period show such remarkable strength and endurance. By making a study of old paper it is possible to tell by the peculiarities of the fibres in what manner they were reduced to pulp. The stampers did not cut or lacerate the delicate fibres, the action of these machines was rather to rub and fray the material.

The papermakers of Holland with their windmills found it difficult to compete with the mills of Germany where there was plenty of water power. The ingenious Dutch therefore tried to devise a method of macerating rags that would require less power

NO. 109

An old German paper mill showing the stampers. This equipment, now in the Deutsches Museum, München, was originally used in a paper mill in Haynsburg, founded 1700 and continuing in operation until 1909

than the stamping-mills and at the same time be more productive than the ancient method. In the latter part of the seventeenth or early eighteenth century, a Dutchman, whose name has been lost to posterity, conceived the idea of the cylinder beater for reducing rags to fibres suitable for the making of paper. This machine is known to this day as the "Hollander," after the country of its invention. It consisted of an oblong wooden tub, rounded at both ends, in which revolved a solid wooden roll made from the trunk of a tree and fitted with about thirty iron knives. The linen and cotton rags circulated around the tub and were lacerated by the action of the metal bars of the roll revolving over a metal or stone bed-plate set in the bottom of the tub directly under the roll. The material was kept in constant motion by the impetus given by a backfall and by the rotation of the roll. With the invention of the Hollander the putrification of the rags ceased in some mills, the cloth being washed and beaten without preliminary treatment.

The Hollander was naturally first used in the country of its origin. It is not known when the invention was taken to Germany, but no doubt it was shortly after Sturm's visit to Holland, which is later referred to. As early as 1725, the papermaker Katerstein in Saxony wrote: "the Hollander in Freiburg furnishes in one day as much pulp as eight stamper-holes do in eight days." This was but seven years after Sturm compiled his book depicting the newly invented beating-engine. Even after the introduction of the Dutch machine many mills in France and Germany, as well as in Italy, retained the old stampers for the breaking of rags, and employed the Hollanders to complete the beating. In 1716 one of the largest paper mills in Germany operated twenty-five sets of stampers, of four each, which supplied three dipping vats with paper stock. This same mill, in 1777, had thirteen sets of stampers and three Hollanders, which macerated sufficient stock for workmen to make paper from four dipping vats. In 1829 this mill reduced the number of stampers to eight sets and increased the Hollanders to four. This equipment furnished the beaten stock to the same four vats. As early as 1760 one of the largest paper mills in France had discarded the stampers altogether, but operated twelve of the Dutch beaters which supplied the liquid stuff for forming the sheets of paper from twenty vats. A mill of this magnitude would have had a capacity of about seventy-five reams of finished paper a day.

The first paper mill in central Massachusetts was established by

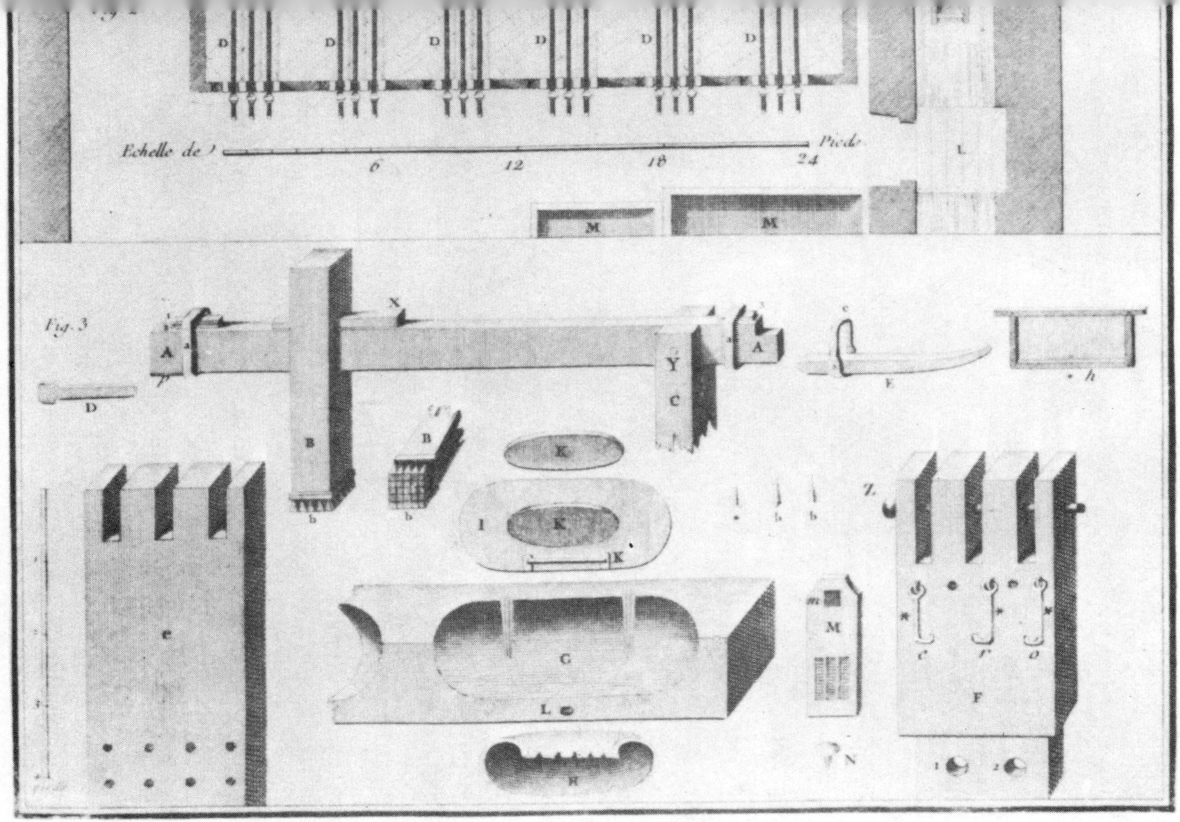

NO. 110

An elaborate stamping-mill as shown in *Art de faire le Papier* by de Lalande

Abijah Burbank in 1775 and was considered at the time one of the best equipped mills in this country. This establishment operated two Hollanders with rolls two feet in length and twenty-six inches in diameter, which produced the required amount of stock for two

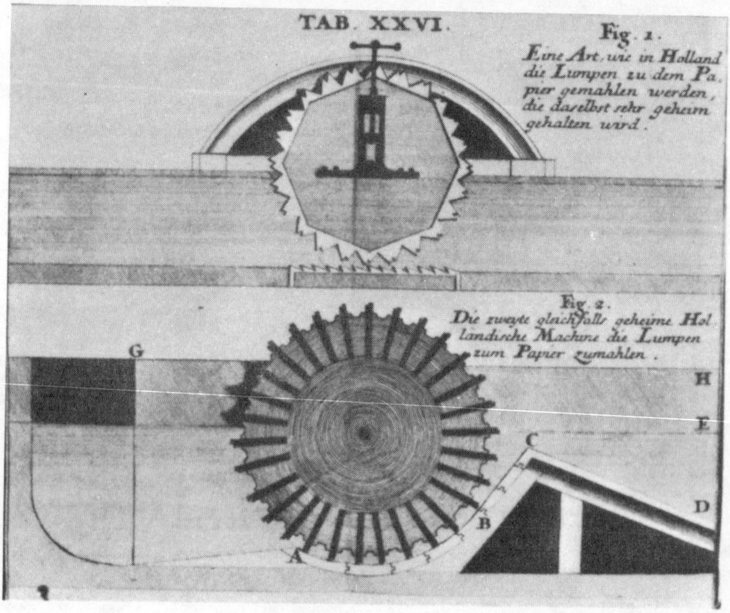

NO. III

The first engraving of the Hollander. From *Vollständige Mühlen Baukunst* by Leonhardt Christoph Sturm, Augsburg, 1718

vats. A mill of this size would have had a capacity of from four to five thousand sheets of demy paper a day.

The earliest engraving of the Dutch beating engine is shown in the book by Leonhardt Christoph Sturm dealing with mill machinery which was printed in Augsburg in 1718, the title being "Vollständige Mühlen Baukunst." In this volume the Hollanders are pictured in a circle, so that four or five of the machines could have been actuated by one huge wooden cog wheel, which in turn

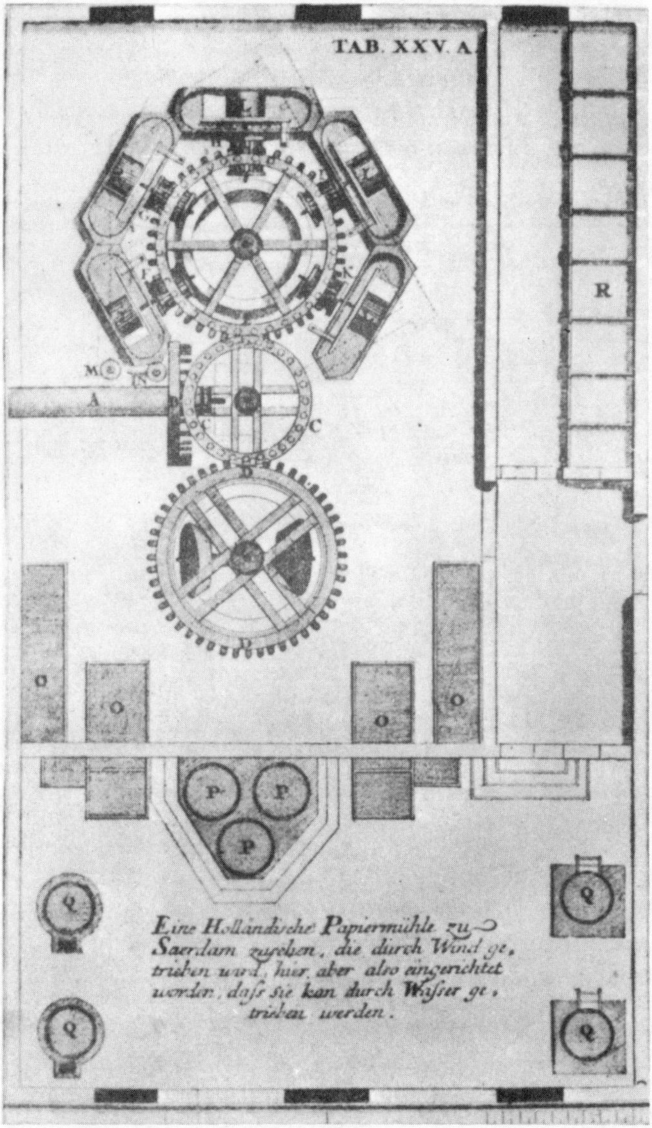

NO. 112

From Sturm's *Vollständige Mühlen Baukunst,* 1718, showing five Hollanders and two mill-stones, all operated by one water wheel

was operated by a windmill or water wheel,—in Holland by the power of the wind, and in Germany and France by the weight and force of water. In one of the prints in Sturm's book there are depicted five Dutch beaters in combination with two millstones, the

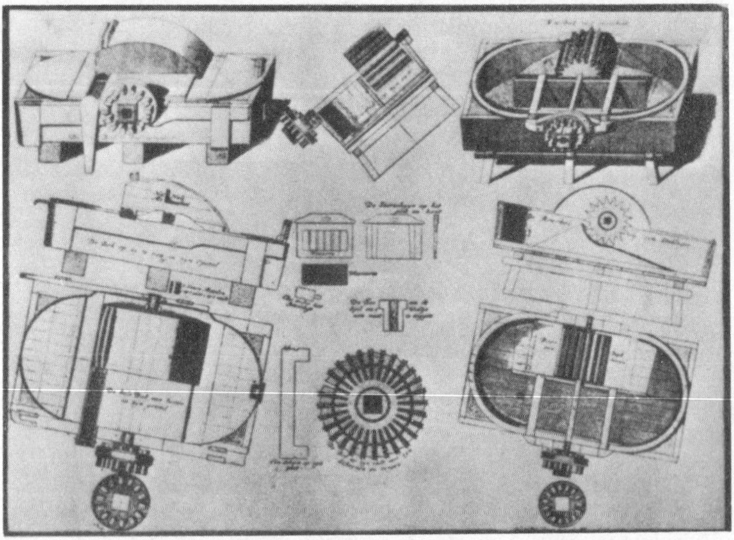

NO. 113

The improved Hollander. The engraving in *Groot Volkomen Moolenboek,* Amsterdam, 1734

stones grinding flour at the same time the beaters reduced rags to pulp for papermaking. This combination of making flour and paper in the same mill was not usually practiced, and some have thought that the placing of the millstones in the engraving was simply a fancy of the author, who produced the book after a journey through Holland visiting mills used for various purposes. It is, however, recorded that Nicholas Hasselbaugh, who was operating the former De Wees mill in Pennsylvania in the late eighteenth century, had a pair of stones in connection with the paper mill and carried on an extensive business in both flour and paper.

Not over fifteen years ago I visited a handmade paper mill in the south of England and saw there in operation a large water wheel the power of which was being used for operating two Hollanders and two sets of millstones, both papermaking and the milling of grain being carried on under the same roof. This mill was demolished about ten years ago, and as I was present at the time I acquired some of the old equipment and brought it to America, where it will eventually be used in the making of paper.

In the Hollanders of the middle eighteenth century the washing and cleansing of the linen and cotton was accomplished by the use of two screens, called "chesses," which extended through the hood, or "chapiter," just in front of the revolving roll. The screen nearest the roll was fashioned of coarsely-woven metal wire and the other of fine cloth made of horsehair. At the base of these two screens a hole-scupper or "dalot" was placed which carried away the unclean water as it was thrown through both of the screens by the velocity of the revolving roll. Fresh water was supplied continually to the beater tub during the washing process and the mission of the two screens was to prevent the escape and waste of any fibrous material. An engraving in *Groot Volkomen Moolenboek,* published in Amsterdam in 1734 depicts this form of washer, and is the first illustration of the Hollander to appear in the country of its invention.

The German papermakers were slow to take advantage of the newly devised Dutch machine, no doubt thinking, and rightly, too, that the old method of macerating the cloth by use of the stamping-mills made the best paper. Both Frederick the Great and Maria Theresa encouraged the German masters to install the Dutch beaters and discard the old-time stampers. In 1800 there was but one paper mill in England where the stampers were in use. The French were the last to adopt Hollanders exclusively, and until 1861 the French Government insisted that all pulp which

A
A
B
B
C
D
F
F
G
f
g
d
d

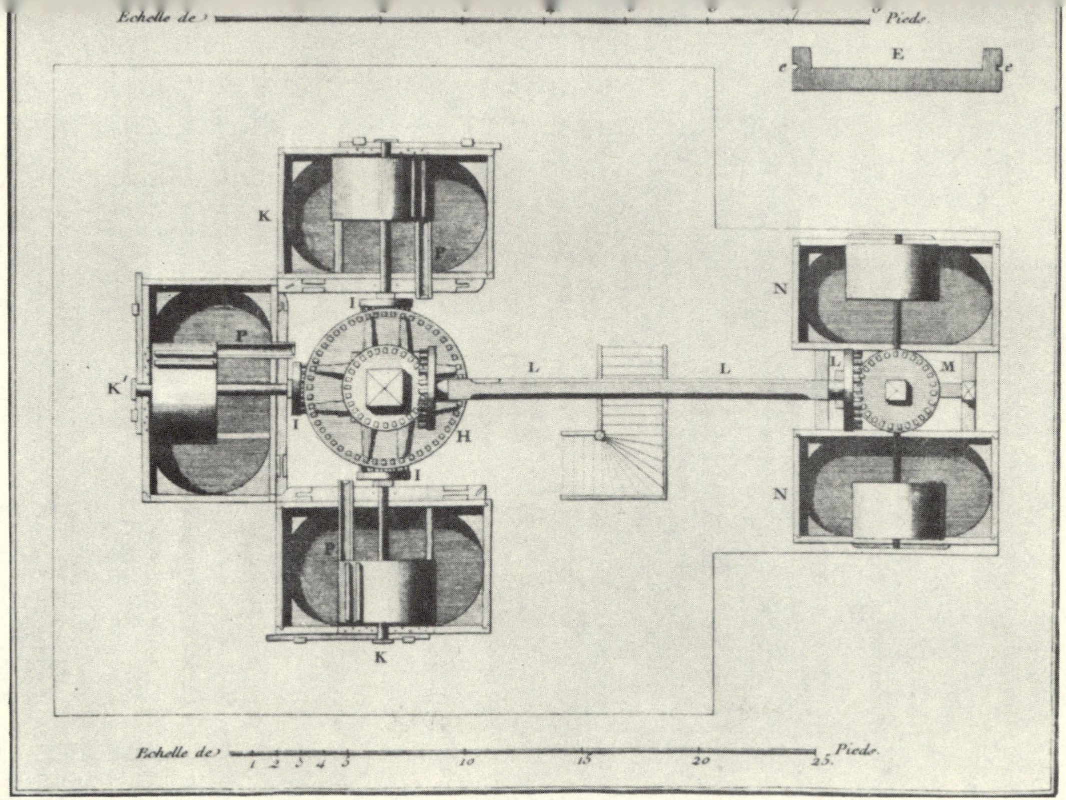

NO. 114

The Hollander of the eighteenth century showing the chesses

was to be used for making paper for the stamp-office be beaten by the primitive stamping-mills,—assuring long-fibred stock and therefore stronger and more durable paper. In 1801 Johann Christoph Ludwig recommended the German papermakers to

NO. 115

The operation of Hollanders by the power of the wind. From the engraving in the rare pamphlet of 1801 by Johann Christoph Ludwig

adopt the Dutch windmills as a means of driving the beating-engines, writing a small pamphlet, now rare, regarding his observations. As late as 1920 there was a set of the old stampers in operation in the South of the Tyrol, and several Italian and German mills still retain them in their buildings, as curiosities. Only recently I was offered a set of these ancient stampers by the last member of one of the old German papermaking families, the troughs of stone worn by use and the wooden parts almost falling into decay.

In the Hollanders of the present day a few improvements have been made, but in general principle and construction they are almost identical with the old machine of Dutch invention shown in Sturm's book of machinery of 1718.

PART IV

PART IV

THE INVENTION AND EVOLUTION OF THE PAPERMAKING MOULD, THE PRINCIPAL TOOL OF THE ANCIENT MAKERS OF PAPER

NO matter at what remote period the art of papermaking had its beginning in China, the mould was from the first the most important tool of the craft, and through the centuries in both the Orient and Occident it remains to this day the chief implement of the makers of handmade paper.

The earliest papermaking mould, like that of the present time, consisted of a square or rectangular framework of wood, not unlike a very shallow box without top or bottom; one surface of the wooden frame being covered with a material,—at first woven cloth, then bamboo strips, and later metal wires—close enough to retain the macerated paper fibre upon its surface, and yet open enough to allow the water to drain away. Across the framework a number of ribs or bars were placed which acted as supports to the mould covering. Many varieties of wood have been used in past centuries for the mould framework, but during the entire history of the craft bamboo, oak, pine, mahogany, pear, and fir have predominated.

The earliest papermakers, the Chinese, probably used a mould made with a bamboo framework over which was stretched a piece of coarsely woven fabric. As the wet sheet of paper could not have been taken from the cloth, the sheet was dried on the mould and then removed. In this earliest form of papermaking a number of moulds would have been necessary if much paper was to be produced, as it would have required at least a day for each sheet to dry before it could have been removed from the cloth covering.

The earliest history of papermaking in the Orient is naturally obscure, but it is thought that an ancient Persian of genius conceived the idea of a mould from which the wet sheet of paper could be removed while still moist. This constituted the first real

NO. 116

It is thought that the first Chinese papermakers used a bamboo frame covered with woven cloth for forming sheets of paper

step in papermaking, as it enabled the artisan to form sheets continually upon the same mould. For this purpose the mould covering had to be constructed of a smooth and firm material from which the moist sheet would readily free itself. Such a mould covering was made by placing thin strips of rounded bamboo side by side and stitching or lacing them together at regular intervals with silk, flax, camel-, yak-, or horsehair. The bamboo strips, as well as the stitches, left impressions in every sheet of paper made upon the mould. The marks or indentations made by the pieces of bamboo

NO. 117

The papermaking moulds of India have changed but little in hundreds of years. This particular example was procured in the village of Kondapalli, near Bezwada. The framework and wedge-shaped ribs are of light-weight wood, while the mould covering is of bamboo, the deckle strips are at either side of the mould

are now known as "laid-lines," while the less noticeable impressions of the stitches are termed "chain-lines," as the hair lacing consisted of chain stitching over and under the strips of bamboo. In a Koran manuscript in Kufic characters, attributed to the twelfth century, now before me, I find by holding the paper to the light that there were in the original mould sixteen bamboo strips to

NO. 118

These four photographs show the manner in which the Bezwada mould is dipped into the vat, while illustration 121 shows the way the sheet is couched, or taken from, the bamboo cover

NO. 119

A Japanese mould and deckle for making sheets of paper about eighteen by twenty inches. There are twenty laid-lines to the inch and the chain-lines, or hair stitches, are placed one and one-half inches apart

every inch, as there are sixteen laid-lines to every inch in the paper. The hair stitching of the mould that was used in forming the paper upon which this Koran was written was so finely executed that the chain-line impressions are not visible in the pages, but the hair lacings, nevertheless, formed part of the construction of the original mould.

The laced bamboo mould cover resembled a piece of matting and when a sheet of paper was to be made, the covering was placed loosely on the shallow wooden framework, which supported it firmly; the entire mould, both frame and cover, was then dipped into the vat of macerated pulp and brought to the surface loaded with the wet, fibrous material. The thin paper stock was prevented from flowing beyond the top and bottom edges of the matting by bamboo rods laced to the mat at its extremities, parallel with the "laid" bamboo strips. The flow of pulp on the right and left sides was kept within boundaries by the workman, who dipped the mould, holding sticks, one in each hand, along the outer edges of the matting, parallel to the "chain" stitches. Thus the four outside edges were fenced in such a way that the wet paper stock or pulp was kept within bounds, forming a sheet of paper almost the size of the mould covering. In couching a sheet of paper the matting was lifted free from the framework and the wet layer of paper deposited flat upon a board, the workman rolling up the matting, from the top edge to the bottom, leaving the moist, tender sheet firm and unwrinkled.

While this type of mould construction seems complicated, it was the simplest and best method that could have been devised, and upon this early Oriental form of mould the manner of making paper in all ages has rested, even the most modern paper-machine employs precisely the same principles. In my collection of moulds for the making of paper there are specimens from China, Japan, Indo-China, Burma, and all parts of India. While these moulds

NO. 120

A mould and deckle from Tonkin, French Indo-China. Each locality in the Orient has its own particular form of mould, but these implements are all formed on the same general principle

may differ in the construction of the framework and the manner of keeping the wet paper stock within bounds, the top covering, or matting, is usually of the same form of fabrication, and the principle of the whole operation of papermaking is almost identical.

NO. 121

Showing the manner in which a sheet of paper is removed from the flexible bamboo covering

These bamboo moulds of the laid pattern were used by the Persians and the Chinese centuries before the introduction of papermaking into Europe. To my knowledge, no watermark in the form of an object or design has ever been found in paper from the period of the vegetable filament moulds. Perhaps the ancients did not think of this means of marking their paper; or perhaps it may have occurred to them, and finding the bamboo strips so inflexible the idea was not carried out. The laid and chain-lines of the moulds caused indentations in the paper unintentionally, but the use of insignia or devices was not practiced by the Oriental workers in paper.

It is not possible to determine the date of the introduction of bamboo moulds, nor can it be stated at what time metal wire succeeded vegetable filaments as a material for mould covering. The Moors established papermaking in Spain in the eleventh or twelfth

NO. 122

The back of a French mould showing the ribs which support the laid wire covering

century, and it is thought that wire replaced the natural fibre moulds in Europe not long after that time. At whatever period wire took the place of bamboo, the metal was used in precisely the same way as the rounded bamboo reeds, and the metal wires continued to impress the laid and chain-lines in every sheet of paper formed upon them, reproducing almost in counterpart the lines left by the bamboo moulds.

There is great variation in the distance between the chain-lines and also in the number of laid-lines to the inch, in sheets of paper made on either the ancient bamboo or the later metal wire moulds. The finest Oriental bamboo mould in my possession was made in

Indo-China and has 38 laid strips to the inch, which would produce this same number of indentations in each sheet of paper made thereon. The coarsest mould covering, or matting, that I have is from Bengal and has but 12 of these rounded bamboo

NO. 123

An Indian mould complete with framework, loose bamboo cover, and two deckle sticks. This mould is from Bengal and was used for making paper about thirty-four inches square

splints to the inch. The greatest number of metal wires to the inch in any European mould in my collection is 32, a French mould of the early nineteenth century; the coarsest counts but 12 to the inch, a sixteenth century Italian mould. The paper used in the Gutenberg Bible has about 28 laid-lines, while some of the paper

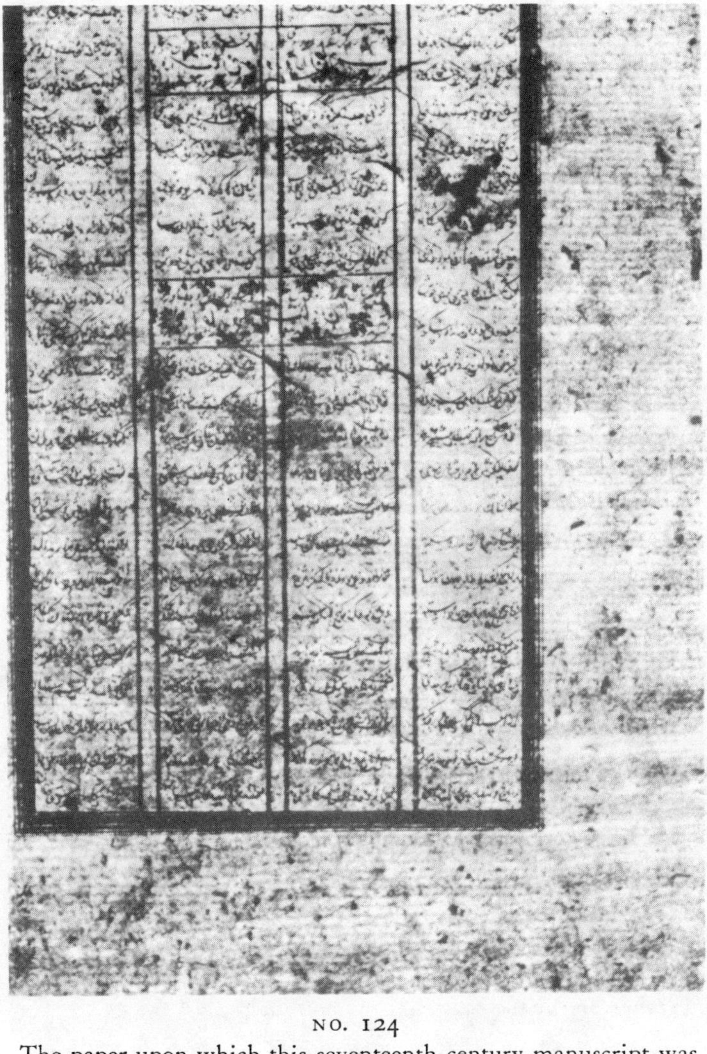

NO. 124

The paper upon which this seventeenth century manuscript was written was fabricated on a bamboo mould with thirteen laid lines to the inch. The manuscript page is from Firdausi's Shahnama, or Book of Kings, said to be the longest poem in the Persian language

used by Gutenberg's successor, Peter Schoeffer, has 24 lines to the inch. Moulds that had 32 wires to the inch were in use at the John Tate mill, the first to be established in England, in 1496. The paper used by Erhart Ratdolt at Augsburg, in 1490–1492, shows from 32

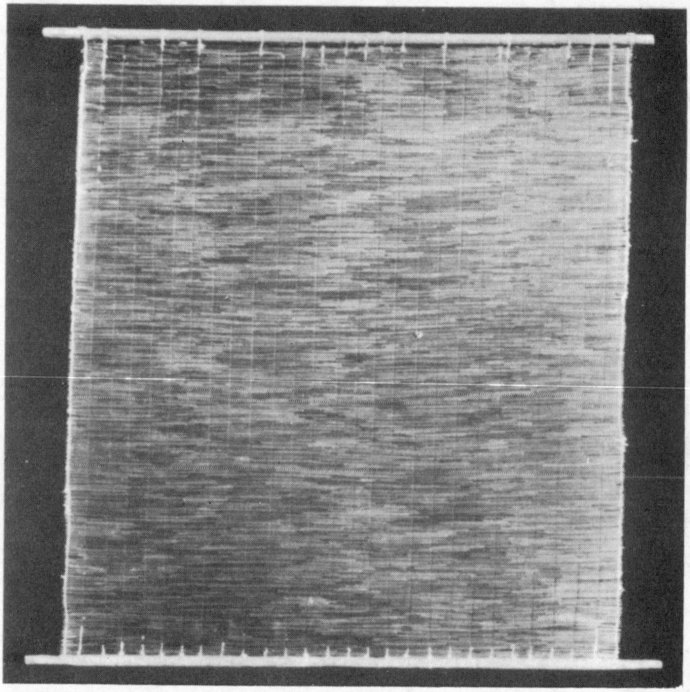

NO. 125

A bamboo mould covering from India. The bamboo splints or reeds which form the laid-lines are of various lengths, all being held together by the chain-lines of horsehair

to 38 lines to the inch, while the paper printed on by Berthold Ruppel in Basle has but twenty. The chain-lines or sewing wires varied greatly also and are found spaced from one half an inch to fully two inches apart. From this it will be seen that location or period has little bearing on the number of lines found in old papers, and that therefore there is no possible way of cataloguing or indexing these impressions in paper so that a particular time or place of origin may be determined. Ancient paper, both Oriental

NO. 126

A bamboo mould covering from Indo-China in which there are thirty-eight laid-lines to the inch

and Occidental, must be placed by its general characteristics, and not by any list of rules that can be given as infallible.

Metal wire was originally made in Europe by beating the material into thin plates, and then cutting it into strips and rounding them with a hammer. The earliest Occidental moulds must have been made with wires fashioned in this manner, as the first watermark dates from about 1270 and was no doubt formed of metal, and not of bamboo. Wire making by use of the draw-plate was probably known in Nürnberg early in the 14th century; the *History of Augsburg,* 1351, and that of Nürnberg, 1360, mention the "wire-drawers" (Drahtzieher). The draw-plate was imported into France by Richard Archal, and first made its appearance in England in 1565, having been taken to that country by a German workman named Schultz. An English patent by Brockedon, dated 1819, specifies the use of diamonds, rubies, sapphires, and other gems, drilled for draw-eyes and mounted in iron, to be used in the drawing of wire. The draw-plate, made of metal or of hard precious stones, like most other appliances thought to have had their origin in Europe, was without doubt first used in the Orient many centuries before it was conceived in the Occident.

In France a company or guild of wiredrawers existed previous to 1583, and in London there was an incorporated company under the title of "The Art and Mystery of Drawing and Flattening of Wire." The motto of these artisans was *Love draws friendship,* and they had their work-shops in Crooked Lane before the alterations in London Bridge. It is thought that iron wire was used first as a mould covering at the introduction of papermaking into Spain, but the use of brass was centuries old at this time and brass soon took the place of iron as a more suitable metal for the purpose.

In European moulds the wires were permanently fastened to the wooden frames, and in this way differed from the Oriental method of using detached covers. The European moulds were rigid, so that the wet sheets of paper formed thereon were couched in a manner different from the Chinese or Persian fashion. This change constituted, in early centuries as at present, the greatest deviation between Oriental and Occidental methods of operation. In forming sheets of paper in the East the paper stock was kept from flowing over the mould by the use of loose sticks, while in the West the wet, fibrous mass was kept within bounds by a fence joined permanently at the corners and placed on the mould when a sheet of paper was to be formed, and removed so that the sheet might be couched. This narrow rim or fence was called a "deckle" and it not only prevented the macerated fibre from flowing over the sides of the mould, but determined the size of the sheet of paper as well. The term deckle, which is now used in the English language, had its origin from either the German word "deckel" or old Dutch "dekfel," meaning cover. As the word was not used in England before the day of John Spilman, the German, who established a paper mill in Dartford, Kent, in 1588, he, in all probability, introduced the term into English paper mills.

Until the latter part of the eighteenth century the chain wires had been laced or sewed directly to the wooden ribs, or supports,

NO. 127

An early nineteenth century French mould and deckle made entirely of oak. There are thirty-two laid wires to every inch

of the moulds, which caused the pulp to lie heavier along each side of every chain-line in the sheets of paper. In *fig. a* this thickness or denseness is unusually well defined, all European paper made before about 1800 may be distinguished by this peculiarity. Paper of this kind is rightly called "antique laid." Modern papermakers do not use this term correctly, but apply it to any rough-surfaced laid paper. Old paper with pronounced irregularities in the laid-lines is known as "mediaeval laid," a specimen is shown in *fig. b*. The heaviness along the ribs or chain-lines, noticeable in both antique and mediaeval laid papers, was eliminated by holding the covering away from the ribs of the mould by the use of wires run-

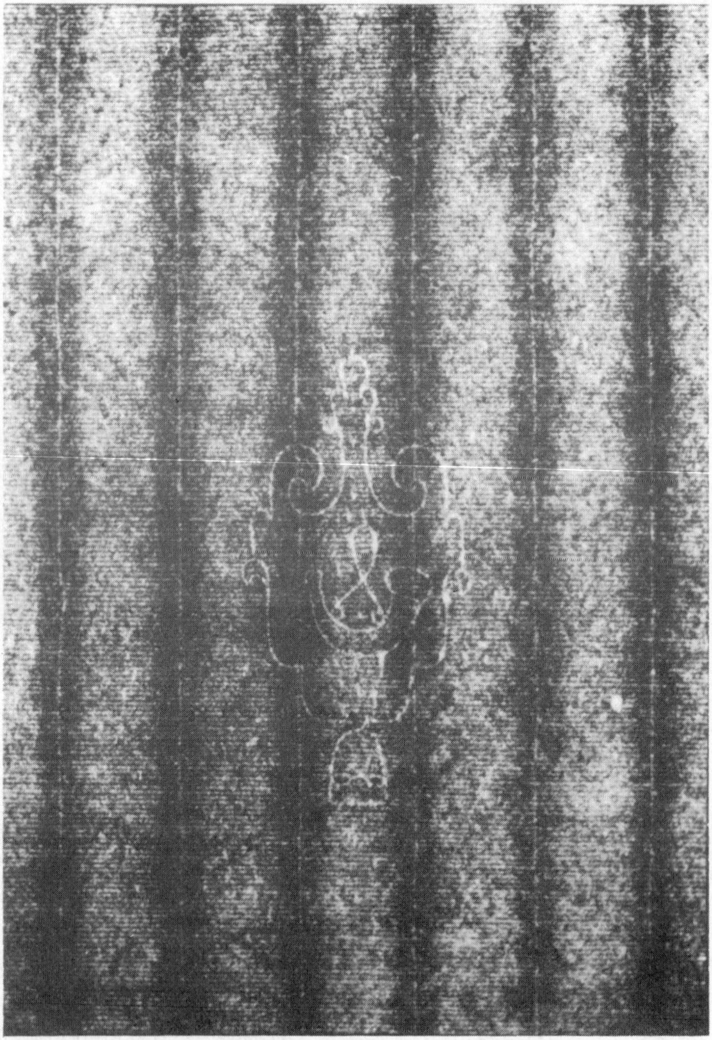

Fig. a

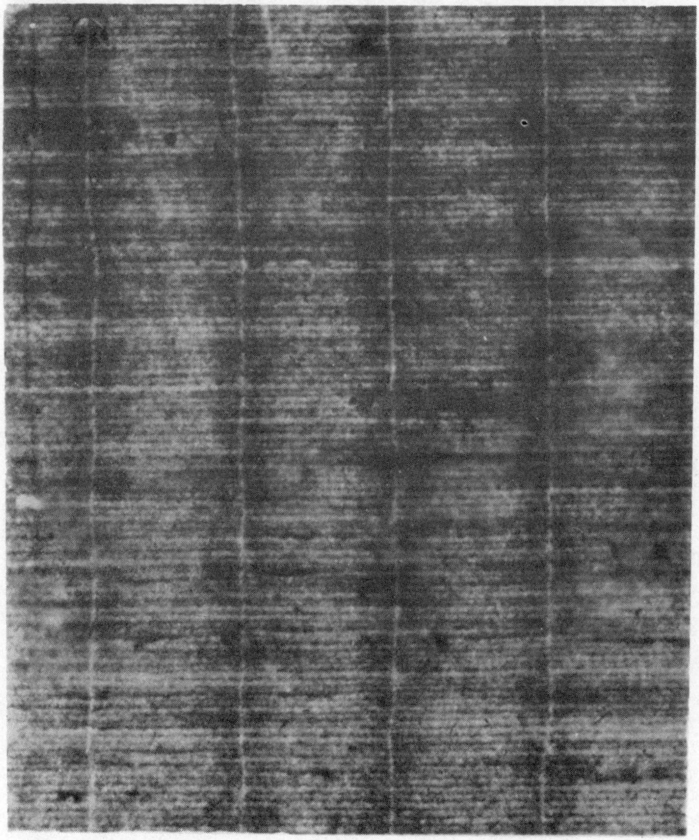

Fig. b

Fig. c

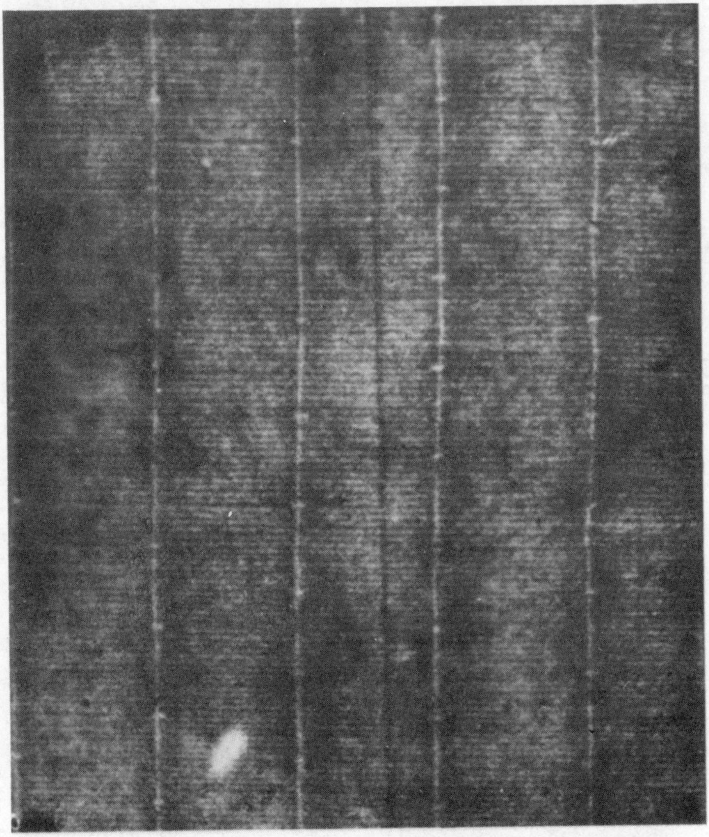

Fig. d

Fig. e

Fig. f

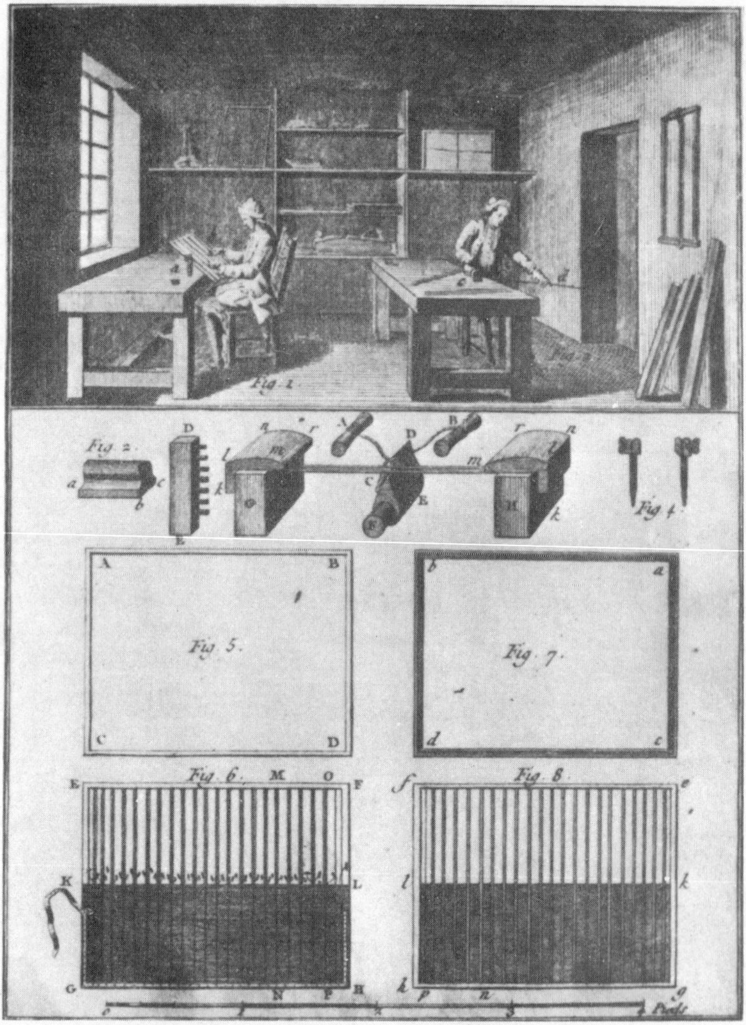

NO. 128

A mould-maker's shop of the eighteenth century with the tools and appliances of the craft. From the engraving in the work on papermaking by Nicolas Desmarets, 1774

ning parallel with, and under, the laid wires. This prevented the paper stock from settling at each side of the chain-lines, a tendency caused by a peculiar suction of the wedge-shaped ribs as the mould was lifted from the vat. The effect produced in the paper by these under wires is shown in *fig. c*. In *figs. d* and *e* the stitching wires which hold the laid wires to the ribs are quite pronounced, while in *fig. f* there are two heavy laid wires running parallel at half-inch intervals. These slight variations are found in great varieties in old paper and give some clue as to period and locality. Very few Oriental moulds had rib supports under each chain-line and therefore the paper made upon them does not show the peculiar characteristics that are apparent in European mediaeval or antique laid papers.

The mould-making required great skill as these implements were constantly in and out of warm water, and if the wooden frames had not been joined properly at the corners they would soon have warped and become useless. When paper was being made, two moulds and one deckle were required at each vat, so it was essential that the deckle fit each mould perfectly. In Oriental papermaking there were two mould coverings and one set of deckle sticks required, with but one wooden frame to support the two bamboo mattings, or mould coverings.

The wood used for European mould-making was prepared by cutting the straight-grained, well-seasoned material into narrow boards, free from all knots and blemishes. These boards were boiled in water, over and over, and finally slowly dried. The process was repeated many times before the material was considered ready for use. The finest present day moulds are each composed of at least three varieties of wood, while the old paper moulds were usually made of but one species.

When the art of papermaking was introduced into the American colonies in the late seventeenth century there were no wire

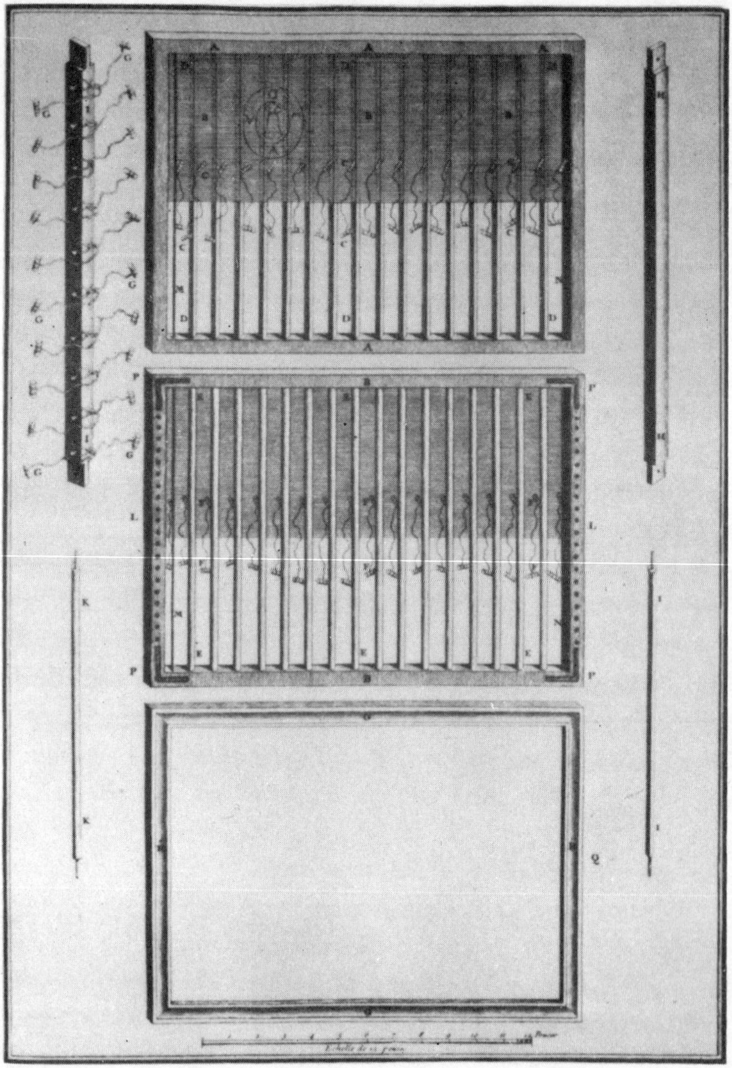

NO. 129

Mould construction as delineated by Joseph Jérôme Le Français de Lalande in *Art de faire le Papier*

drawers or wire workers in this country, so that the few moulds required by the pioneer mill were imported from Europe, possibly from Holland. It was not many years, however, until moulds were being fashioned in their entirety in this country. The first American

NO. 130

The wooden deckle prevented the wet paper stock from flowing over the edges of the mould, and determined the size of the sheet of paper

mould maker was a resident of Germantown, not far distant from the Rittenhouse paper mill. It is not known whether Isaac Langle learned the craft of drawing and weaving wire in Germany, or acquired the trade so that he might supply the Pennsylvania papermakers with the much-needed moulds. In any event he was making moulds for the Ephrata mill in Lancaster County as early as 1740. Aside from this little is known of this pioneer craftsman. After Langle's death in 1744 there appeared an advertisement for the sale of a *Siebmacherswebersthul mit einen dazu gehorigen eisern Schienen-zug und anders zugehar*. (A wire-weavers' frame, with an iron apparatus for wire-drawing, and other belongings). These were the simple and unpretentious tools which had been used by this early paper mould-maker in all of his work.

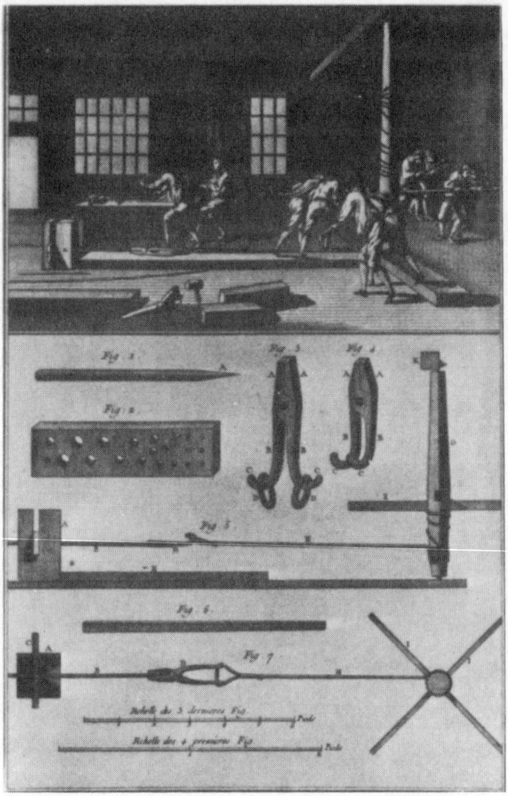

NO. 131

A wire-drawers' shop and the tools of the trade. From a middle eighteenth century French work on the art of metal working

Nathan Sellers, who is usually given the credit of having been the pioneer worker in paper moulds, did not commence his operations until about 1770. A few years later he abandoned his calling so that he could join the Continental army, but his work was considered of such importance to the papermakers of the country that he was requested to return home and resume his trade. Sellers was an expert artisan and executed considerable work for the govern-

NO. 132
The large mould, with deckle in place, is of the antique laid style, while the small mould, without deckle, is of the modern laid type. The difference in the wire coverings can be noticed

ment, many of his moulds were used for forming sheets of paper at the Willcox mill in Pennsylvania where much of the government paper was fabricated. Many of the tools which had been made and used by Sellers, comprising straightening-boards, wire-drawing blocks of lignum-vitae wood, wire-plates, and pincers, were in the possession of one of his descendants until his removal to the West in 1841. One of the most important implements em-

NO. 133

An early specimen of modern laid paper showing the absence of the dark streaks at either side of the chain-lines. This sheet was made about 1800 by J. C. C. Fleisen, Germany, and shows the watermark of this firm

ployed by wire workers was a straightening-board. Sellers had fashioned one of these, after his own ideas and inclinations, which embodied the principles of the ones then used in France and England, but much improved, so much so as to have been re-invented

NO. 134

Most of the Oriental moulds did not have a support under each chain line as did the first wire moulds of Europe

and patented in France as late as 1800. The tool consisted of a series of stout wood or metal pins securely fixed in a board; the wire being drawn between this row of pegs was first bent in one direction, then in the reverse, in a waving line, the bends gradually diminishing or shortening until the wire became perfectly straight. Sellers greatly improved on this appliance by the use of adjustable pins, as late as 1833 his method of working was employed in England and was known there as the "Sellers' Straightener." Nathan Sellers died in 1830, at the age of eighty years. In an account book of 1781 to 1784, written by Robert Morris, Superintendent of

NO. 135
A Dutch antique laid mould with watermarking wires. The deckle is in place ready for a sheet of paper to be made

Finances, the Government records having paid Nathan Sellers on March 12, 1783, the sum of $103.30 for a pair of moulds for papermaking. This same year, owing to the demand for papermaking moulds, Mark Willcox of "Ivy Mills," Chester, Pennsylvania, was obliged to import some moulds from England. The following bill for them is still preserved:

LONDON, *August 2d* 1783

Captn Falconer Bot of Peter Wynne		£	s	d
1 Pair Double Fools Cap Paper Moulds		5	5	
1 Pair Double Post *Do*		5		
Box for *Do*	at 5		2	
2 fine Brass wire washers 24 Inch by 20 is 6–8/		1	13	
2 Pieces coarse *Do* for back wires 6.8	at 1.9		11	8
		£12	12	0

An original bill from Nathan Sellers for the making of moulds is also in existence and reads:

TENCH FRANCIS ESQ. TO NN & DD SELLERS DR

June 18th 1791			
for a fine paper mould for Bank paper with box	£4.	10.	0
for 88 watermark letters at 1.10½	8.	5.	0
for 14 *do* figures		15.	0
for 8 *do* private		8.	0

Recd Paymt Oct 5. 1791
Nn & Dd Sellers

The "wove" mould covering was originated in Europe by John Baskerville (1706–1775), and the date of his rediscovery is given as about 1750. Baskerville was a printer living in Birmingham, England, and his effort was to produce paper for his book printing which should afford a smoother surface than paper of the laid pattern in use in Europe ever since the introduction of the craft into Spain in the twelfth century. The wove covering was made of fine brass screening and received its name because it was woven on a loom in about the same manner as cloth. It left in the paper an indistinct impression resembling a fabric. Baskerville had been in the japanning and metal-working trades before becoming a printer, so that he was naturally familiar with this material, metal screening having been used in England for other purposes a number of years before it was put to use as a material upon which to mould sheets of paper. While this printer of Birmingham was the first European to adopt the wove type of mould, paper of this kind had been fabricated in the Orient for many years, the Orientals having used coarsely woven cloth as a means of retaining the macerated fibre which, when dry, was paper. The first book to be printed on European wove paper was Baskerville's magnificent quarto edi-

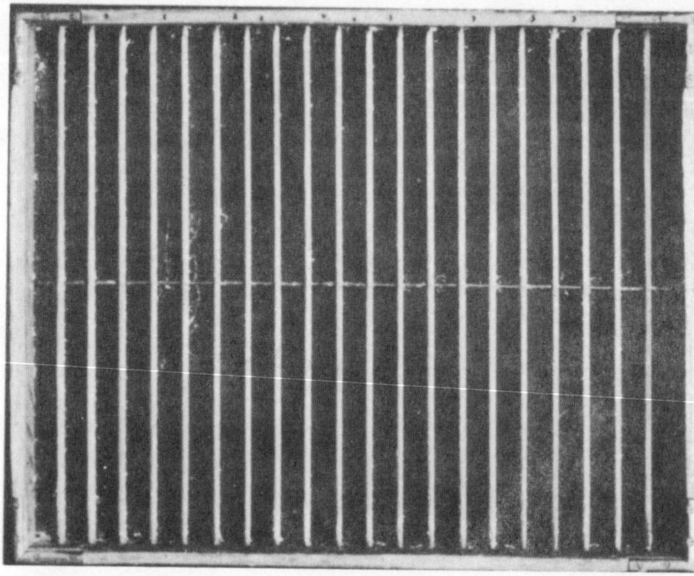

NO. 136
The back of the Dutch antique mould showing the rib supports

tion of *Publii Virgilii Maronis Bucolica, Georgica, et Aeneis,* which was published in Birmingham in 1757. Another edition of Virgil was issued by the same printer in 1771, under the earlier date of 1757. It is the first printing of the 1757 edition that is of interest on account of the paper upon which it was printed.[1] The 1757 edition of Virgil immediately created a furor in English and continental literary circles and Baskerville received no end of praise and

1 The first edition of this work may be distinguished by the fact that the tenth and eleventh books of the *Aeneid* are headed, *Liber Decimus. Aeneidos,* and *Liber Undecimus. Aeneidos,* while in the 1771 printing they appear uniform with the other titles, *Aeneidos Liber Decimus, Aeneidus Liber Undecimus.*

commendation for his work; the new style of smooth paper was acclaimed universally. William Shenstone, the noted English poet, wrote of this work in 1757: "My neighbour, Baskerville, at the close of this month (March, 1757) will publish his fine edition of

NO. 137
A wire-drawer of the middle eighteenth century

Virgil; it will for type and paper be a perfect curiosity." The earliest use of the wove paper known to Rhys Jenkins was in *Philosophical Transactions* for 1788. This writer on the papermaking history of England also states that the publication "Archaeologia" changed from laid to wove paper in 1796. Baskerville's Virgil of 1757 was however the first book in which wove paper was used, and as late as 1790 paper of this kind was common in England. In the copy of Virgil before me there are but 186 of the entire 432 pages on wove paper, the remaining pages (246) being on very fine antique laid paper. In the first wove or *papier vélin* sheets the ribs of the moulds appear distinct and pronounced when the paper is held to the light. This defect was caused by the woven wire mould covering having been sewed directly to the ribs, or supports, of the mould. Perhaps the wove cover that was used in making the paper for the Virgil had been stretched directly over a mould of the laid type. These imperfections appeared in paper until about 1800 when they were eliminated by placing a coarsely-woven metal screen under the top wove covering. It is not known whether Baskerville made his own paper or had it fabricated in a neighbouring mill. In a letter to Horace Walpole, fourth earl of Orford, (1717–1797), written in 1762, Baskerville wrote: "The ink, presses, chases, moulds for casting, and all apparatus for printing were made in my own shops," but he did not state that he possessed any papermaking appliances. From this it may be assumed that he did not operate a paper mill, and it is reasonable to believe that he was kept occupied with the type founding and printing without adding to his labours the making of paper.

In the preface of Baskerville's edition of *Paradise Lost,* 1758, the printer expressed his gratification that the "improvement" in the paper which had been used in the Virgil in the preceding year had been recognised by the public: "I must own," he wrote, "it gives me great satisfaction to find that my edition of Virgil has been so favourably received. The improvement in the manufacture of the paper, the colour, and firmness of the ink were not overlooked." Although the wove paper was acclaimed by Baskerville and the literary public in general, it does not appear that the Birmingham printer made exclusive use of paper of this type, for a number of editions appeared after 1757 imprinted upon paper of the old style, even the Baskerville folio Bible of 1769 is printed entirely upon antique laid paper.

It was wove paper of English fabrication that Benjamin Franklin exhibited in Paris about the year 1777. The novel paper made

NO. 138

A corner section of a wove mould of the type used in making the paper upon which Baskerville's Virgil of 1757 was printed

a favourable impression upon the papermakers and printers of France, and they desired to procure moulds for duplicating this type of paper which did not show any laid lines and which had a smooth surface like a piece of woven silk. In 1779 M. Didot, the noted Parisian printer and publisher, having seen the *papier vélin* that Baskerville had used, addressed a letter to M. Johannot of Annonay, a skilled papermaker, asking him to endeavour to duplicate the smooth and even surface of this new paper. Johannot was successful in his experiments, and for his work in this field he was in 1781 awarded a gold medal by Louis XVI. In 1782 M. Rêveillon produced finely-executed specimens of wove paper, a number of samples of which he presented to the Académie des Sciences, receiving considerable praise for his initial attempts.

The exact date of the introduction of wove moulds in America is not known, but it is thought to have been about the year 1776. In a letter written by George Escol Sellers, the grandson of Nathan Sellers, previously referred to, the following is said regarding moulds of this type: "When in England in 1832, William Matthews, son-in-law and successor of Fausett, I think was the name, mentioned some entries of a shipment of both wire for 'laid,' and 'woven' wire for vellum moulds, to John Sellers, the father of Nathan, as far back as 1773. They were the only weavers of brass vellum (wove) wire at the time, and the principal mould makers in England. Their wire works were at Chepstone, on the River Wye." From this it would appear that wove paper was made in America as early, if not earlier, than in France.

It is thought that the term wove in connection with papermaking was first used in the patent granted to Henry Fourdrinier and dated July 24th, 1806 (NO. 2951). This specification reads in part: "The method of making a machine for manufacturing paper of an indefinite length, laid and *wove* with separate moulds . . . a number of moulds of the description called laid or *wove,* are hooked or fastened together to form one long mould." This term is again mentioned in the English patent of Ledger Didot, dated June 2nd 1812 (NO. 3568), where the specification begins: "Certain other improvements upon the said machines for making both *wove* and laid paper, etc." The definition given in Rees' Cyclopaedia, 1819, is as follows: "By a modern improvement, these marks (laid- and chain-lines) are avoided, and the paper has a smooth, even surface. For this purpose the wire is wove in a loom, exactly like cloth, and stretched over the frame of the mould . . . The wove paper, as it is called when made on these moulds, is a very superior article to the old paper, particularly for books; but a prej-

NO. 139
John Baskerville

udice still prevails in favour of the old paper with lines, which obliges manufacturers still to make it, though by no means so fine or good as wove."

John Baskerville was born in 1706, in Wolverley, Worcestershire. In the year 1745 he entered the japanning trade where a liking for things of a mechanical nature was developed. While still occupied at his first calling he started in the printing trade and in this work became one of the most renowned craftsmen in the annals of English typography. He cut the punches for his type and did much to elevate type design in England. The books printed by him are known for their neatness, brilliancy of ink, and the smoothness of the paper. This he accomplished by placing each printed sheet as it came from the press, while the paper was still damp and the ink undried, between hot plates of polished copper and subjecting them to a slight pressure. Baskerville died in 1775 and from his will, which had been written in 1773, it would appear that he had a mill of some kind on his estate in Birmingham, for it directs that his body is "to be buried in a conical building in my own premises heretofore used as a mill, which I have lately raised higher and painted, and in a vault which I have prepared for it." It may be that Baskerville carried on the craft of papermaking, as several writers suggest, in the building mentioned in his will, but it is probable that this structure was not of sufficient size for this purpose. According to Thomas Hansard in his *Typographia,* 1825, the mill mentioned in the Baskerville will was nothing more than a tomb of masonry in the shape of a cone, under a windmill. Baskerville's house was destroyed in the riots of 1791, but his remains continued undisturbed until 1821, when a canal was cut through the estate and his leaden coffin was exhumed, and reburied in a spot that is now a question of debate.

I do not care for wove handmade paper for book printing, unless it be of Oriental origin; the European prototype lacks character and possesses not nearly the quality and beauty of the original laid paper. It is hardly proper to use old-style type in printing a book upon wove paper, and it is an anachronism to mark machine-made paper with laid-lines, for paper formed on a machine is naturally of the wove variety and any laid-line watermarking applied by use of a dandy-roll borders on deception.

In 1790, John Phipps, an Englishman, patented a process for teaching writing and drawing by means of watermarked lines in paper. The specification is dated August 21st 1790, and reads: "A method to facilitate the acquirement of several of the useful and

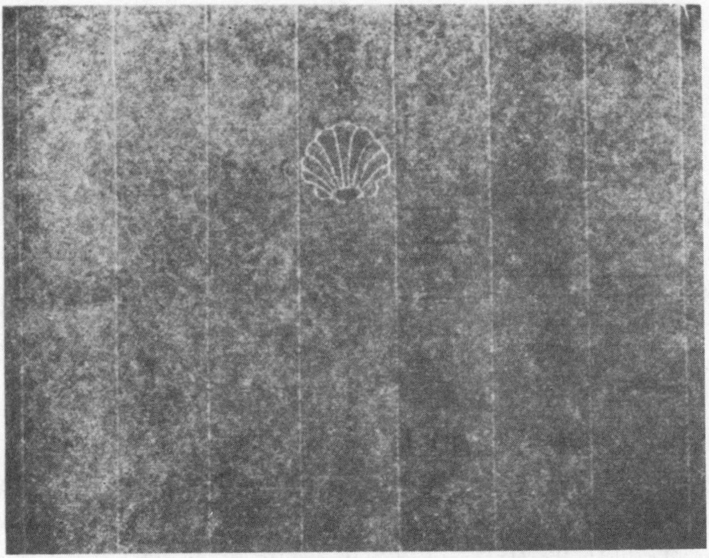

NO. 140
An early modern laid paper with extremely fine laid-lines

polite arts by an easy, effectual, and expeditious manner of teaching writing and drawing, which is done by fabricating the moulds, wove or washing wires, or other wire on which sheets of paper are made, so that every sheet shall come out of or from them properly adapted for the purposes, when finished by the maker in the usual manner, by having the lines made and the copies set for writing, and the outlines and sketch for drawing, by what is called the watermark, for the learner to trace over." This absurd method for the teaching of writing and drawing was but little used, and the interest in the novel idea lies chiefly in the fact that this was the first English patent pertaining to moulds for papermaking. The earliest American patent relating to paper moulds was granted April 11th, 1793, to John Carnes, Jr., of Delaware, the original patent was destroyed by fire in the patent office and no record of it is available. The first English patent to deal with the actual making of moulds followed the American patent by about three months, being dated June 27th, 1793. The abridged description reads: "A machine or engine for working and binding of wire, and making of moulds used by papermakers, and for the more correct and expeditious method of working and binding wire for sieves, screens, etc."

While there are no rules for distinguishing the paper of one century or country from another, it is possible by making a study of the texture and fibre of the paper and of the laid- and chain-lines and the character of their impressions, to arrive at the approximate date and determine in what country and under what conditions the paper was made. With wove paper it is not easy to attach a date to the sheets, but paper of this type was usually watermarked with the year in which it was made, and as wire-wove paper of the Baskerville style has been in use only since about 1750 there is not sufficient age to excite much curiosity from an historical point of view. As the first paper made in the Orient was formed on crudely made moulds covered with loosely woven cloth, we must accept the wove as the original type of paper. The laid bamboo mould came into use a few centuries later, and as it was the first style of mould from which a sheet of paper could be taken while wet, this invention must, as I have said, be considered the most important step in the development of Oriental papermaking. Long after the advent of the bamboo laid mould came the wire laid mould of Europe,—first fashioned with iron wire and later with the more durable and non-rusting brass. In the middle eighteenth century the *vélin* type of mould was re-invented, but instead of using the woven fabric of the East, from which a moist sheet of paper could not be taken, a woven brass screen was employed which formed a firm and rigid surface. At the close of the eighteenth and the beginning of the nineteenth centuries came inventions for making paper by machine. These used the original idea of the transfer, or Oriental, mould, except that in the machine the mould was continuous, and formed the paper, not in single sheets as by hand, but in any desired length. The great paper industry of today is built upon the original Oriental bamboo mould which came into being almost two thousand years ago.

PART V

PART V

THE ECCENTRICITIES OF THE OLD PAPERMAKERS AND THE CHARACTERISTICS OF THE PAPER THEY FABRICATED

THE paper fabricated in the fifteenth and sixteenth century mills did not vary in size to the same extent as paper made in later years, owing to the limited demand for sheets of divers dimensions and the expense of producing them. For every size of paper, two moulds and one deckle were required, appliances fashioned only with difficulty and considerable outlay of money. In the early years of printing, the paper was not cut and the sheets were printed upon in almost the original sizes formed in the moulds, though as a rule the deckle edges were trimmed away. In all antique paper there is considerable variation in the thickness and finish and even in single books the leaves will vary in weight to a marked degree. Usually, the early paper was well formed in the moulds, with very little variation in thickness throughout each individual sheet, which testifies to the skill and adeptness of the vatman. The tone of the old paper was never entirely uniform, and owing to the absence of chemicals in the manufacture the grades of paper differed strikingly in colour; the best paper was of a creamy tint, while the poorer grades, made from old and discoloured materials, were a light coffee tone, and at times even a dark grey. The bleaching of linen and cotton cloth for use in papermaking was not in general use until the late eighteenth century, and all paper made before that time assumed the same tone of the material from which it was made. It is interesting to note that in England very little paper except that of a coarse brown tint was produced until the late seventeenth century, about the time when papermaking was introduced into the American colonies.

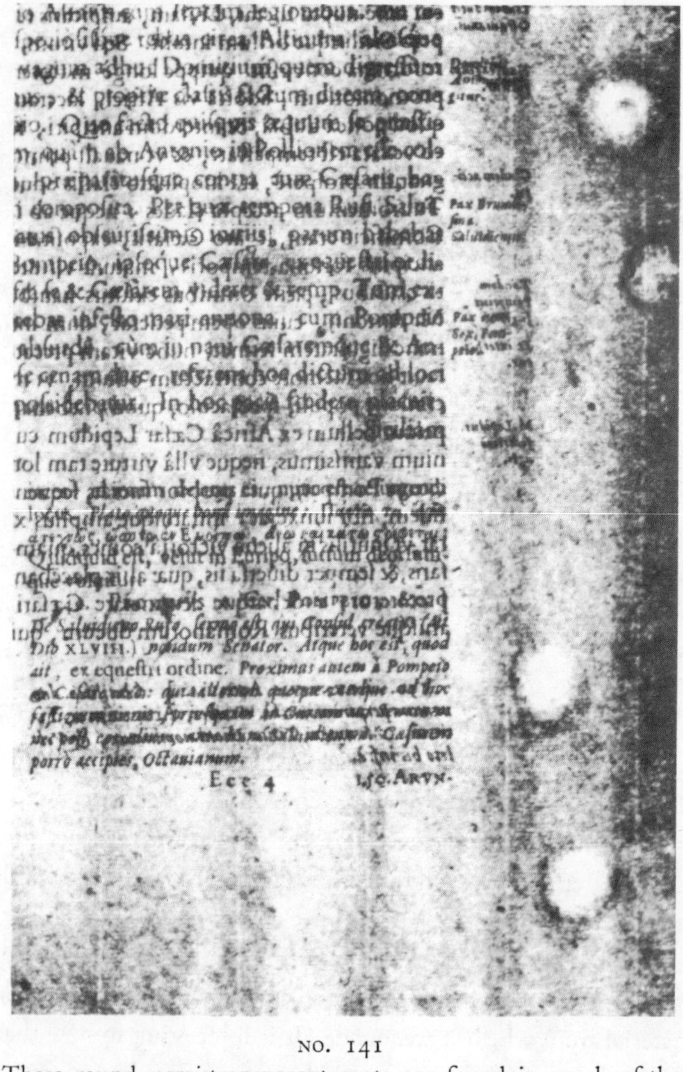

NO. 141

These round, semi-transparent spots are found in much of the old paper and were caused by drops of water falling on the newly-formed sheet. This specimen is of Dutch origin, the printing is from the year 1668

In holding old paper to the light it is not uncommon to notice in the sheets round semi-transparent spots, somewhat smaller than a dime. These spots have a thinner surface than the rest of the sheet, resembling a watermark to some extent, and may be found in

NO. 142

Ruffled or blurred chain-lines are not uncommon in old paper

modern handmade paper as well as in the old. They were caused by drops of water falling from the worker's hands upon the moist sheet just after it had been dipped; the water spreading the wet fibres formed almost perfectly round indentations in the finished paper, discernible only when the sheet is held to the light. When these drops appear in paper there are usually several of them, of various sizes, as a rule near the corners of the sheets. There are also in some of the old papers, ruffled or blurred chain-lines, a slight imperfection not entirely absent in handmade papers of the present day. This blurring of either the chain or laid wire impressions was caused by the coucher having let slip the mould, with its thin deposit of pulp, just as he was in the act of couching the sheet upon the felting. Another defect, encountered only in old paper, is a peppered appearance throughout the sheet, due to the knotting of certain of the fibres. These three peculiarities so common in old paper may be seen in the accompanying photographs. It would

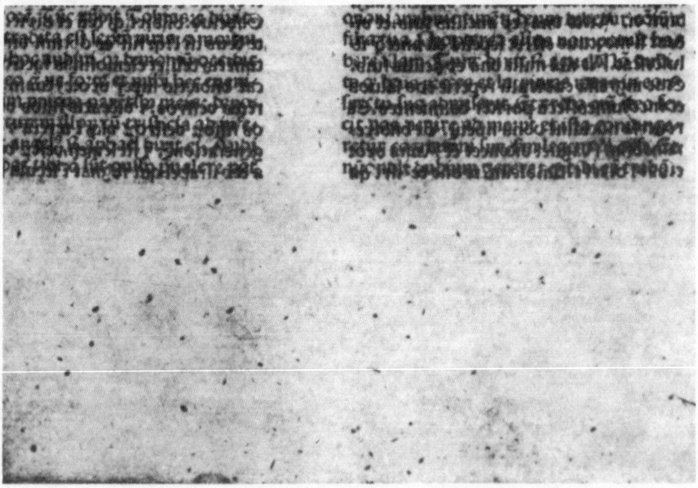

NO. 143

Knotted fibres readily found their way into many early sheets of paper. This particular example is of German origin, printed upon by Anton Koberger, Nürnberg, 1478

hardly be possible to examine thoroughly any fifteenth or sixteenth century book without finding all of these imperfections represented in the paper. It is not unusual to discover imbedded in old papers hairs loosened from the felt in the pressing process; or human hair fallen from the head of the vatman or coucher. In my collection there is a piece of paper in which has been imbedded a fifteenth century mosquito, visible only when the sheet is held to the light. To prevent the intermingling of foreign substances with the paper pulp was a trying problem for the old papermakers, even at the present time the artisans have no end of trouble with stray

NO. 144

The rough edges on old paper were caused by the deckle, or cover, which fitted around the edges of the mould. In a. is shown an early American paper with extremely rough edges, caused by a badly fitted deckle. The edges shown in b. and c. are more characteristic of old deckles

particles which enter the beater, and, broken there into hundreds of minute pieces, find their way into the vat and cause unsightly specks in the sheets of paper.

In examining early printed books it will be noticed that in many instances the deckle or rough edges have been cut away. In some cases this trimming was done before the paper was printed upon, as many of the old typographers no doubt regarded the deckle edges simply as necessary imperfections in the making of the paper. When the deckle edges were left undisturbed by the printer they often suffered at the hands of the binder, so it is more usual to find early volumes with cut edges than with the original deckles. The weapon with which the binder dealt the most deadly blows was the "plow," with which he cut away the margins, placing the printing in a false position to the page of the book, and often de-

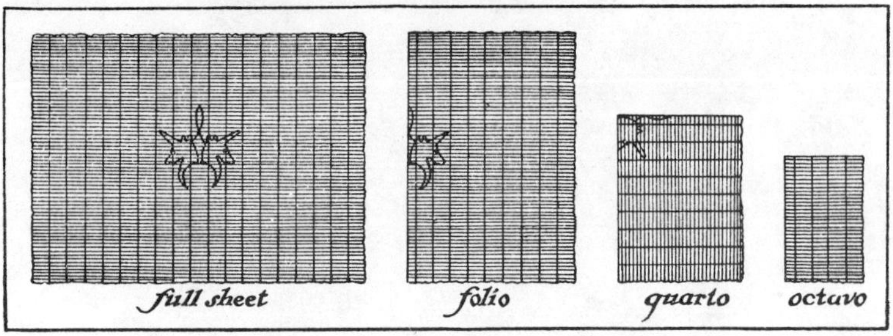

NO. 145

In a folio volume the sheets were folded in the centre, the chain-lines running up and down

nuding the volumes of a portion of the very text. In binding, books were often trimmed to such an extent that a folio became a quarto, or a quarto an octavo, in size. Modern bibliographers usually catalogue these old volumes by measurement instead of by the paper. If a book was printed on a once-folded sheet, it would always be a folio, no matter how much a ruthless or ignorant binder may have cut it down. In the early printed books this trimming was possible as the margins were purposely left wide for the notes of the readers, so that a folio could have been reduced several inches without interfering with the text. The only way to arrive at the original size of an old volume is to examine the paper on which it was printed. In the paper moulds the chain-lines invariably ran the short way of the sheet, and the laid-lines the long way, or from right to left as the mould was held in the hands of the vatman. In binding a folio volume the sheets were folded in the centre, the chain-lines running up and down, parallel with the fold. No matter if much of the margins have been trimmed off, it is always possible by this means to see that the volume was originally a folio. In a quarto the paper was again folded so that the chain-lines ran across the page, from left to right. With an octavo the chain-lines ran up and down as in a folio.

Before the writer is a book from the Aldine Press dated 1522 which was catalogued as an octavo, for its actual measurements (12 x 20 cm.) would indicate this size. Originally, however, the volume was a quarto, as the chain-lines in the paper run across the pages. On the other hand, the edition of *Artemus de Bello Gallico,* printed by Nicolas Jenson in Venice in 1471, is usually because of its large size, set down by the cataloguers as a folio, but in this book the chain-lines run across the pages, so it must be a quarto. This test holds good for a great many books, between octavo and folio, it can be relied upon because it is a natural test, and furthermore it is applicable to the books of all Occidental nationalities.

The size of the largest paper used by William Caxton was 22 by 15¾ inches. This paper was used for two books only,—the first and second editions of the *Golden Legend* (first edition finished at Westminster the 20th of November, 1483; the second edition appearing soon after, but without date. Printed in double columns, containing 464 folios). This size was probably found too large for convenient printing, as for all of the other Caxton books smaller paper was used, varying in size from 16 by 11 to 18½ by 13 inches. The average size of moulds used for making paper in Europe during the fifteenth century was about 19 by 14 inches, while the largest sheets measured about 26½ by 18½ inches. These were convenient sizes for the mouldmakers to fabricate and also for the papermakers to form sheets upon; the larger the moulds the more difficult they were to handle, as their weight when laden with the wet pulp was far greater than one would think. In Great Britain, even as late as 1818, it was ordained by law that no newspaper should exceed in dimensions 32 by 22 inches. "So desirous were the publishers of newspapers at that time for a larger surface," wrote Charles Cowan, in his reminiscences of his early days in a handmade-paper mill at St. Mary Cray, Kent, "that an order for

NO. 146
A mould for making two sheets of paper at one dipping. This type of divided mould was first used about 1826

a large quantity of paper reached us, the size to be 32½ by 22½ inches, and the stationer stated that he would assume all risk of passing it through the Stamp Office." This would have given a surface of 731 square inches to each sheet instead of the statutory maximum of 704 square inches. About the largest surface of the fifteenth century papers was 470 square inches, while a surface of approximately 235 square inches was the most common. The largest size of handmade paper that is fabricated at present is made in the James Whatman, Springfield mill at Maidstone, Kent, the sheet being termed "antiquarian" and measuring 53 by 31 inches, a surface of 1643 square inches! For making sheets of this giant size the mould is hung on a mechanical lifting device and six or eight men are required in the dipping and couching processes. It was at this renowned Kentish mill that the size of handmade paper known as "medium" (22 by 17½ inches), was first made two sheets on a mould, a deckle running through the mould dividing it into two equal parts so that two sheets could be dipped and couched with practically the same effort as had been previously expended in making one sheet. This was in 1826, and the innovation caused a split among the makers belonging to "the Original Society" which lasted several years. The section of artisans who were against making two sheets of medium paper simultaneously on a mould, called themselves the "Deckles," while those agreeing to make the paper in the new and novel manner were known as the "Stars." After a reconciliation of the two factions, Mr. Balston, the owner of the Whatman mill, summoned William Grigsby, one of his workers and the leader of the Stars, who always wore on his jacket a ten pointed star, denoting the number of vats in the mill, and gave him a sovereign for each point, bidding the vatman and couchers spend the money in a jollification.

The Oriental papermakers have never had any standard sizes for paper, and throughout Japan, China, India, and Indo-China the dimensions of the moulds vary greatly. The largest Japanese mould in my possession would be capable of producing sheets of paper 66 by 30 inches (1980 square inches), the bamboo laid strips or reeds running lengthwise of the mould, as in European metal wire-covered moulds. This exceptionally large mould was procured near Okayama, a town in South-western Japan, and had been used for making a low grade of "hanshi" paper. Paper made upon a mould of this size was usually cut into sheets about twelve inches square. In Japan, as well as in China, sheets of paper have always been fabricated in almost every conceivable size, but in late years, due to the exportation of paper from these countries, the paper has been to a great extent standardized. All through India, however, the sizes and methods remain much as they were hundreds of years ago, as no paper is exported there is little use for uniformity, the various small mills fabricating the paper for local markets only. In the Southern Shan States the largest size moulds

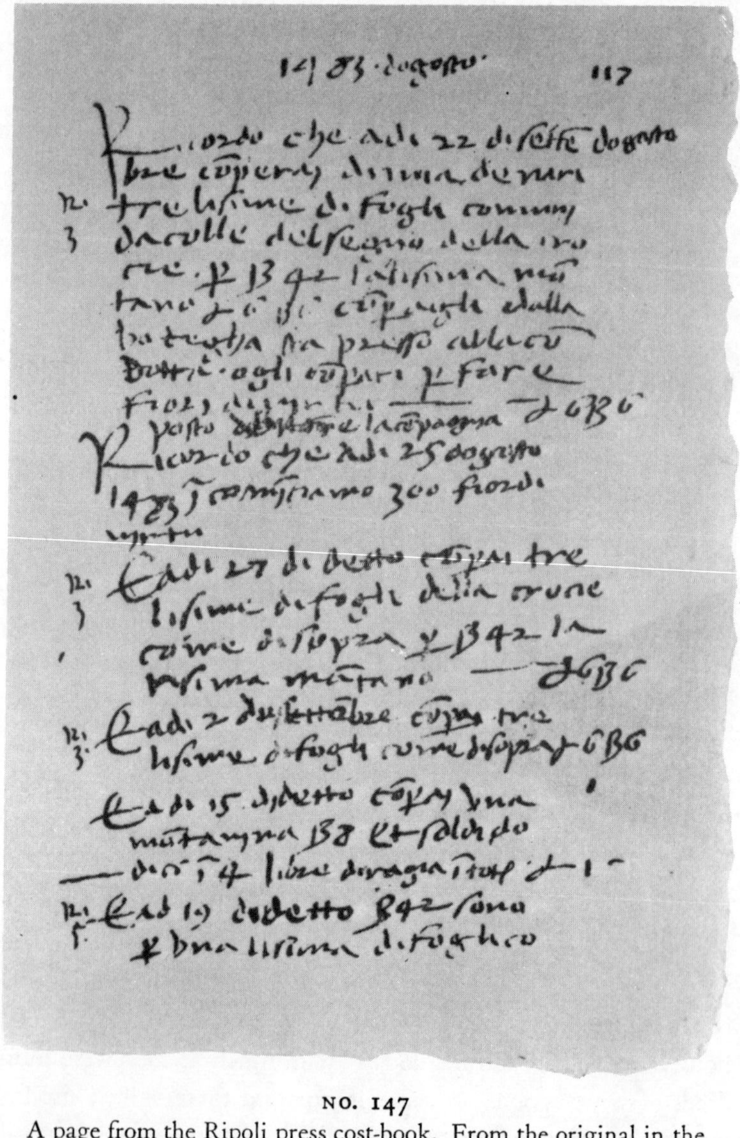

NO. 147

A page from the Ripoli press cost-book. From the original in the Biblioteca Nazionale, Florence

measure 60 by 24 inches and are of the wove type, being covered with woven cloth. This large size of paper is produced east of Taunggyi at Möng Kung, and the paper is termed "kalats." These large sheets, like the Japanese "hanshi," are cut to suit the various purposes of the native population. The Shan paper is made from the paper mulberry (Broussonetia Papyrifera), called locally "hsai-le" or "hsa-le." In Kashmir and Jammu the sheets of paper produced measure about 34 by 34 inches, a complete surface of 1156 square inches. During the sixteenth and seventeenth centuries there was very fine paper made in Kashmir from bamboo pulp. At that time the paper was fabricated in Kagazpura (literally "the paper village") near Daulatābād (Deogiri) in the Deccan. This paper was called "barsāhā." The paper of Kashmir is now made at Khaghzigahi, near Srinagar. In Bengal the paper usually measures about 22½ by 21½ inches in the uncut sheets, while in Tonkin, French Indo-China, where papermaking has been carried on undisturbed for hundreds of years, the moulds are smaller, averaging about 26½ by 11½ inches. The sizes of the Oriental papers have been cited for comparison with the old European sheets; the Oriental papers have not changed in size through the centuries to the extent of the European handmade papers.

Some idea of the cost of European paper in the fifteenth century may be gleaned from the prices that were paid by the directors of the Ripoli Press at Florence, between 1474 and 1483. An original cost-book of this establishment is preserved in the Biblioteca Nazionale in the city of its origin. This book shows that the nine sizes or qualities of paper would have varied in price from about two dollars a ream for the lesser size or quality to about six dollars a ream for the Bologna paper in common folio. The oldest bill that is recorded showing the price of paper is dated 1352 and reads: "To George Cosyn, for one quartern of royal paper, to make painters' patterns, 10d." According to this, a quire of royal paper in 1352 sold for about one-third less than it fetched in the year 1483. In the contract between William Rittenhouse and William Bradford, dated 1697, the prices of paper were given, "ye writing paper at 20 shillings (the ream) and ye brown paper at 6 shillings pr. ream."

NO. 148

A paper mill of the middle eighteenth century, from a contemporary engraving

The weekly wage of a papermaker in England during the latter part of the fifteenth century was from two to three shillings. When the prices paid for paper at this time are considered, the wage of a workman seems low, but at that time labour did not hold the same important place that it did in later years. During the eighteenth and nineteenth centuries the selling price of a ream of finished paper would have offset the weekly wage of one papermaker. In England, during the early part of the eighteenth century, a vatman earned from fifteen to eighteen shillings a week, the wage in America at that time being a little higher than in the papermaking districts of Southern England. Eighteen shillings a week was a high wage compared with other trades, as a bookbinder at that time received but twelve shillings a week and was required to work from six in the morning until nine at night, three hours longer than a papermaker's day. The following interesting account appeared in a little book of trades in 1747 and gives some idea of the requirements at that time of both the master and

NO. 149
The interior of an early eighteenth century German paper mill

the worker: "Papermaking requires much water and a great deal of room, and therefore is altogether carried on at water-mills in the country, which undertakings are not numerous. It goes through various operations and divers hands before it is completed, and the moulding part, which is the principal, requires a nice hand and good eye. They take with an apprentice 5 to 10 pounds who work from six to six, and he ought not to be a very tender lad; they pay a journeyman moulder (vatman) 15 to 18 shillings a week; and a mill with the proper utensils (besides which there must be a reserve of at least 200 pounds, cash for a stock of rags, etc.) will stand one, minded to be a master, in 100 pounds, of which some have two or three. It is a very curious art taken as a whole, and so useful a manufacture, that it ought to be encouraged at home more than it is." From this account it will be seen that the life of a worker in an eighteenth century paper mill was anything but indolent. He was required to be at the mill by six in the morning ready for work, and did not leave for his home until six in the evening. The noon-day lunch was usually consumed while the men were working, but the more lenient masters allowed fifteen or twenty minutes for the repast. From the English book of trades of 1747 it would appear that papermakers at that time were looked upon with some degree of favour, but a hundred years earlier such was evidently not the case according to a petition dated 1636 to the Privy Council from the people of Buckinghamshire and Middlesex. The people living near the paper mills alleged that by converting corn mills into mills for the making of paper the rents were advanced greatly and that the papermakers brought many poor and indigent to the parish whom they had to maintain. Also they complained that the papermakers were too highly paid and were able to save more from their wages than was possible with workers in other trades. The people further alleged that the papermakers brought plague into the country from their rags, that they killed the fish by their double water wheels and by penning up the water flooded the countryside. They also petitioned that the paper was poorly made and would bear ink only on one side and was sold much dearer than formerly.

Many of the old German and French mills were provided with kitchens and sleeping rooms which were used by some of the papermakers, others living in small cottages near their work. The papermaking centres were little communities in themselves, and many of the old artisans lived for one generation after another in the same spot, working at the same vat and with the same moulds that had been in use over long periods of time. In Holland it was the custom for some of the workers to live with the master of the mill where they were working, all having their meals at the same

NO. 150

The trade mark label of a Braunschweig papermaker, "Andreas Bernhart Papiermacher auff der Ocker." A figure of a horse was also used as a watermark by this German papermaker. This specimen is from the first half of the sixteenth century

table. From time immemorial it had been considered proper for the papermakers to cease eating when the master of the house, by laying down his spoon, gave evidence that he had finished his repast. There is a tradition that many of the early Dutch masters were anything but liberal, and that in order to prevent their workers from consuming too much food, the dishes were served steaming hot, and before they had had time to cool sufficiently to be eaten, the master would lay down his spoon and the poor paper-

NO. 151

A label used by Johann Mieser for marking the packages of paper from his mill. German, 1600

makers would be forced to leave their master's table before their lusty appetites had been satisfied. To this day, in the old papermaking districts of Holland, whenever the children are noticed to be gulping their food they are chided by their parents, who ask them if they have been reared by the papermakers. In the summer time the old Dutch artisans began their work in the mills at three-thirty in the morning and ended their tasks at ten, the remainder of the day being spent in the fields and gardens, where they continued the constant stooping which was essential in the making of

NO. 152

The paper label of André Caprony à la Serra, papier fin au Raisin, French, 1661–1669

paper. The old workers, especially the vatmen and couchers, had to have strong and robust constitutions, but the constant stooping posture, combined with the heat of the paper stock in the vat, caused them to grow old prematurely, and at fifty many of these hard working craftsmen appeared to have reached the allotted three score years and ten. During the eighteenth century a great deal of paper was produced in Holland, it being recorded that in the year 1726 there were sold at Amsterdam a hundred thousand reams in a single day. William Rittenhouse, the first papermaker in the province of Pennsylvania, had worked in Amsterdam as early as 1678, when he was thirty-four years of age, but he had learned the craft of papermaking in his native city, Mülheim on

NO. 153

Wood-block for printing a paper label. "Claude Caprony La Serra, papier fin à la Croix blanche, 1689."

the Ruhr. Rittenhouse was forty-four when he came to Pennsylvania in 1688 and it is interesting to surmise just how he had been influenced by the Dutch papermakers and to what extent he introduced the old Dutch customs into the pioneer paper mill of the New World.

The old European papermakers, as well as the early American workers at this craft, were much too fond of their drink and could be found almost any night at the public houses which were located near the mills. The vatman and coucher, who interchanged their work, could readily have been distinguished by their red, muscular arms and hands, and by their stooping backs. This condition was brought about by these workers having their arms in and out of warm water constantly, and by continually bending over the dipping-vat and coucher's-stool. The old mills were inadequately

NO. 154
A French paper label, dated 1707

heated, and the vat-houses, where the moulding of the paper took place, were usually located where the light was poor, the atmosphere damp and gloomy, and the whole environment unhealthy. All through the history of papermaking by hand there was a scarcity of workers, due no doubt to the unwillingness of apprentices to learn a trade which was so disagreeable and arduous; only those with exceptional physical endurance could have remained long at the craft. The greatest strain was required of the vatman, owing to the four-way motion that he had to give the mould in forming

NO. 155
A Dutch paper label, early nineteenth century

each separate sheet of paper. After working years at the vat the craftsmen sometimes lost this "stroke" and were never again able to form a sheet of paper; sometimes this paralysis has been known to attack workmen for different periods during which time they were unable to work at the moulding of paper, then the ability to give the "stroke" or "shake" returned. Yet with all the disagreeable features of their craft, the old papermakers were a happy and contented guild of workmen. As in other callings, there were those who disliked the monotony of living and working in one place, and these journeyed from one mill to another obtaining work wherever they might go. The papermakers were extremely proud of their craft, for the apprenticeship was long, the work demanded dexterity and skill, and to become a proficient moulder and coucher of fine paper required no little ability of head and hand.

NO. 156
A paper label of German origin, late eighteenth century

Nothing remains today of the old handmade paper industry in America, as here, though not in the Old World, the advent of the papermaking machine soon swept aside the ancient, tedious method of forming each sheet of paper separately. Just at the close of the Civil War all papermaking by hand had ceased on the North American continent. Most of the buildings which housed this important early industry have gone into decay, and not a vestige of any of the old equipment, except a few scattered moulds, remains. The original Rittenhouse mill that was established near Germantown, Pennsylvania, about 1690 was swept away by high water in

NO. 157

Wood-block used in the printing of labels for the marking of packages of paper. Chambéry, 1727

1700, but Rittenhouse and his son Claus built another mill, a short distance below the site of the first, which was in operation in 1702. There is no authoritative picture of the original mill, nor of the second structure. According to the letters of William McCulloch to Isaiah Thomas written between 1812 and 1815, the original Rittenhouse mill was of logs, the building extending over an undershot water wheel, there being no dam or race. This Philadelphia antiquarian also states in his letters: "The Rittenhouses, one or more of them, have continued papermakers to this day. The first stone mill is still attached to the present mill and covers a large press with a screw, made of cherry, about eighteen inches diameter in the eye. This press verily appears to be an ancient affair. The mill is not at present employed by a Rittenhouse. It is rented as a paper mill; and the present papermaker Nicholas Rittenhouse employs another mill in the vicinity . . . Jacob Rittenhouse made the large wooden screw, which was about the year 1755 for his uncle William. Jacob, the millwright, also invented, about the year 1760, the slanting plates for grindings rags; as straight ones only, with much delay of grinding, were in use before this discovery . . . Jacob is still living, he is now 86 years of age, is blind, but retains an excellent memory unimpaired."

The buildings usually pictured as the Rittenhouse paper mill are two stone houses that are still standing. The larger of these structures was built in 1707 and was originally the home of William Rittenhouse and his son Claus, there being carved in the stone above one of the windows the inscription $\frac{\text{C}\ \overset{\text{W}}{-}\ \text{R}}{1707}$. This picturesque old stone house, with its steep, narrow, winding stairway and small cheerful rooms, is always pictured as the Rittenhouse paper mill, but was used only as a residence, never as a mill. To the public at large this old two-story structure is better known today as the birthplace of David Rittenhouse, the astronomer (1732–1796), than as the home of William Rittenhouse, the pioneer papermaker. It may be said without hesitation that nothing remains of the actual Rittenhouse mill, either the building or the equipment. Indeed, even the spot where this mill stood is unknown, nor will it ever be possible to trace with certainty the exact location. A like fate has overtaken the William De Wees mill, the second paper mill to be set up in the colonies. Of the buildings that housed it nothing remains. This mill was built in 1710, and was a direct outgrowth of the Rittenhouse establishment, erected near the pioneer mill, on the west side of the Wissahickon Creek on that part of Germantown known at that period as Crefield.

The third mill, also in the province of Pennsylvania, was established by Thomas Willcox, about 1729, on the west branch of Chester Creek in that part of Chester County which in later years became Delaware County. It is recorded that as late as 1884 some of the buildings of this mill were standing, but whether or not

NO. 158
The post-horn label was common throughout the nineteenth century. This example is of Dutch origin, dating from 1813

these were part of the original structure it is not possible to say. When I visited the site of this old mill a number of years ago nothing remained but the stone foundations, which stood in soggy ground almost covered with rank vegetation. A remnant of the old ivy vine, the original slip of which was brought to this country by Thomas Willcox from Devonshire in 1725, still climbed over the ruined walls. It was because of the abundant growth of this vine that the mill in its prosperous days was known as "Ivy Mills," and carried the device of an ivy leaf as its watermark. At the base of one of the crumbling masonry walls, deeply imbedded in the marshy ground, lay a rusty and broken screw press fallen upon its side. It had been one of the indispensable implements of the old mill, and between its platens ream after ream of paper had been pressed,—paper that had been made for ante-bellum banks, for the counting-houses of some of America's oldest firms, and for the banks of the South American Republics as well as those of Greece. It was not until March, 1866 that the mill abandoned the manufacture of handmade paper. On the hill above the mill stands the old Willcox home, still occupied by a direct descendant of this noted papermaking family. While the present occupant is not engaged in the craft of his ancestors, he has inherited an interest and love for papermaking, and many drawers were opened so that I might see specimens of papers fabricated at Ivy Mills generations ago. Some of these early American papers are now in my possession and are most highly prized. The old Willcox family burying ground, enclosed within an iron fence, lies up the hill, behind the house, and here rest this illustrious family of papermakers. Within the enclosure, but to one side, a small grave stone with the one word CAESAR marks the tomb of an old negro slave whom Thomas Willcox had purchased from the captain of a slave-trading ship that had anchored in Delaware Bay. Caesar had been a faithful and intelligent papermaker in the Willcox mill, and at his death he was privileged to share the resting place of his masters.

About 1736 a paper mill was established on the banks of the Cocalico Creek, near Ephrata, Lancaster County, Pennsylvania, by a branch of the Pietists of Germany, who came to this country in the early eighteenth century. This was the fourth paper mill in Pennsylvania. When I visited the site of this mill a half score years ago there was nothing of the old buildings remaining, not even the foundations could be traced, for a woolen factory had been built upon the spot where the mill once stood. The building usually pictured in histories as the original *papiermühle* of this German religious colony was never a mill of any sort, but the home of one of the leaders of the cult. Not the oldest citizens of Ephrata could remember seeing the paper mill, it had been de-

NO. 159
An American label, middle nineteenth century

molished many years before their days; no authentic picture exists.

Another locality visited, with the hope of finding at least a remnant of an early American paper mill, was the site of the first papermaking establishment in New York, which was set up in 1773 on what is now Silver Lake, a peaceful body of water that flows into the head of Hempstead Harbour, at Roslyn, Long Island. This enterprise was founded by Hendrick Onderdonk and two associates encouraged thereto by Hugh Gaine, a book-seller and printer of New York. The mill was prosperous and Onderdonk became a wealthy and influential man, living in a correspondingly handsome style. Whenever a noted personage came to the village the master of the paper mill was invariably called upon to entertain him, as Onderdonk was the leader in the small community. On the morning of April 24th, 1790, word came suddenly to the village, and to the Onderdonk household in particular, that the President of the United States was on his way through the island, and that he would stop with the paper-mill owner for breakfast. The Onderdonk family was around the table eating roasted clams when the message was received, and a great commotion arose among the women to get things in readiness for the honoured guest. They sprang from their benches, and in haste swept into their aprons the clam shells and cups and saucers, soon having the table cleared for a more formal repast for General Washington. After breakfast the President and Hendrick Onderdonk strolled to the hill west of the pond, and the General remarked that the locality would make an excellent place for a fort. They then visited the paper mill, the noted guest seeming much interested in the process of forming sheets of paper and asking many questions in regard to the methods employed. He was much absorbed in watching the men dip and couch the thin sheets, one at a time, and the vatman summoned sufficient courage politely to request that the President try his hand at the work. This Washington did, dipping the laid mould into the vat and bringing it to the surface laden with the snowy-white fibrous liquid; he then couched the sheet, the workman taking care to slip a small bit of paper between the felts which covered the sheet the visitor had made. Perhaps this means of identification was later found unnecessary, as the paper formed by a novice is generally distinguishable by its own characteristics. It is said that this sheet was a cherished possession of the mill for many years, but no trace of it can now be found. General Washington was evidently much impressed by his visit to the Onderdonk mill, for in his diary under the date of his sojourn there he had this to say: "I left Mr. Young's, Oyster Bay, before 6 o'clock on Saturday morning and

NO. 160
An American paper wrapper of a hundred years ago

passing through Mosquito Cove, breakfasted at a Mr. Onderdonk's at the head of a little bay where we were kindly received and well entertained. This gentleman works a grist and two paper mills, the last of which he seems to carry on with spirit and profit." The old stone foundations of the Onderdonk mill are all that now remain of New York's first paper mill; within the low foundation a great tree grows which clearly shows that it is long since the wooden mill fell into ruin.

Of the old American paper mills, the best preserved structure is one that was erected in the late eighteenth century. It stands in Burnside, Connecticut, on the outskirts of Hartford, looking most pitiful amid its shabby modern surroundings. Originally it stood in the country, beside a stream, but houses and factories have pressed closer and closer, until now the fine old structure must be demolished. The impressive building is about 40 by 30 feet, two stories in height, and built of red sand-stone, each block beautifully squared, and all laid together faultlessly in massive walls; by contrast the neighbouring present-day buildings look mean and poor. The first floor of this old stone mill is divided into two rooms, one served as the vat-house, the other for the beating of rags. The second floor was used for drying and finishing; the original sliding shutters, which had been used to introduce or exclude the air for the drying of paper, are still intact and in fairly good working order. By going over this small building it is possible to picture every phase of the old hand process and to determine just how the artisans operated. The limitations of light and space and the crude appliances did not seem to deter the early workers, and the paper they fabricated, not only served its purpose at the time, but much of it has lasted throughout the years and given to us the records of the past.

In writing about these few early American mills it has not been my desire to give complete historical accounts of them, this has

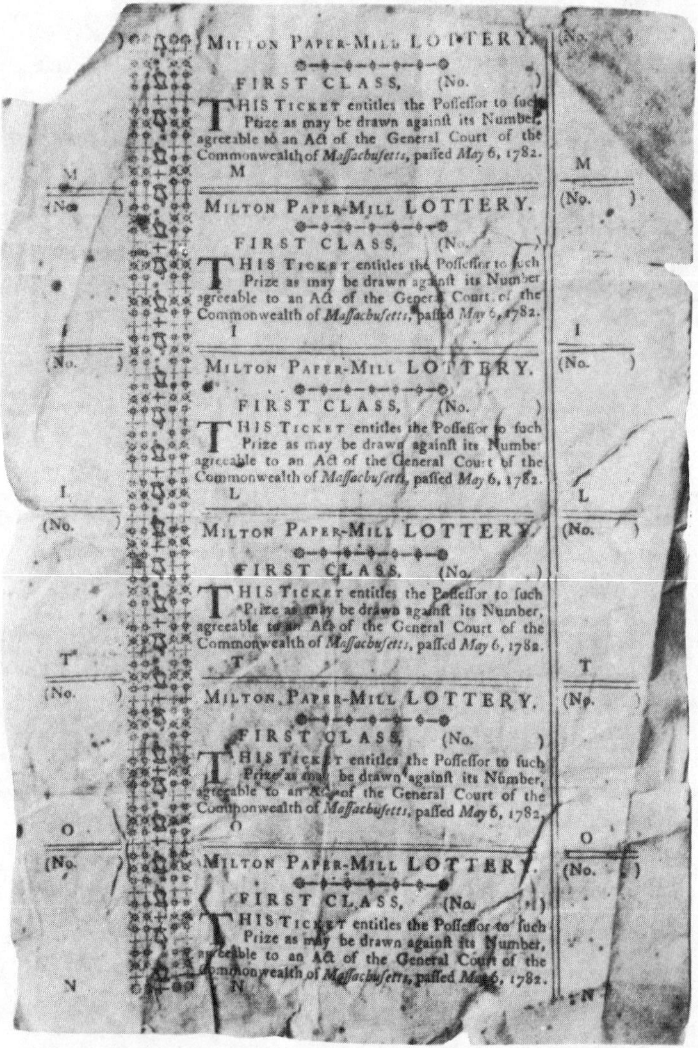

(No.) MILTON PAPER-MILL LOTTERY. (No.)

FIRST CLASS, (No.)

THIS TICKET entitles the Possessor to such Prize as may be drawn against its Number, agreeable to an Act of the General Court of the Commonwealth of *Massachusetts*, passed *May* 6, 1782.

M M M

(No.) MILTON PAPER-MILL LOTTERY. (No.)

FIRST CLASS, (No.)

THIS TICKET entitles the Possessor to such Prize as may be drawn against its Number, agreeable to an Act of the General Court of the Commonwealth of *Massachusetts*, passed *May* 6, 1782.

I I I

(No.) MILTON PAPER-MILL LOTTERY. (No.)

FIRST CLASS, (No.)

THIS TICKET entitles the Possessor to such Prize as may be drawn against its Number, agreeable to an Act of the General Court of the Commonwealth of *Massachusetts*, passed *May* 6, 1782.

L L L

(No.) MILTON PAPER-MILL LOTTERY. (No.)

FIRST CLASS, (No.)

THIS TICKET entitles the Possessor to such Prize as may be drawn against its Number, agreeable to an Act of the General Court of the Commonwealth of *Massachusetts*, passed *May* 6, 1782.

T T T

(No.) MILTON PAPER-MILL LOTTERY. (No.)

FIRST CLASS, (No.)

THIS TICKET entitles the Possessor to such Prize as may be drawn against its Number, agreeable to an Act of the General Court of the Commonwealth of *Massachusetts*, passed *May* 6, 1782.

O O O

(No.) MILTON PAPER-MILL LOTTERY. (No.)

FIRST CLASS, (No.)

THIS TICKET entitles the Possessor to such Prize as may be drawn against its Number, agreeable to an Act of the General Court of the Commonwealth of *Massachusetts*, passed *May* 6, 1782.

N N N

NO. 161

Lottery tickets of Milton, Massachusetts paper mill. Money was raised in this manner for various purposes and even for the establishment of paper mills. This example is from the late eighteenth century

been ably done in other volumes; my thought has been to set down brief descriptions of these mills as they now exist, so that a student of the history of papermaking in this country will know what to expect when he visits the sites of these pioneer paper mills.

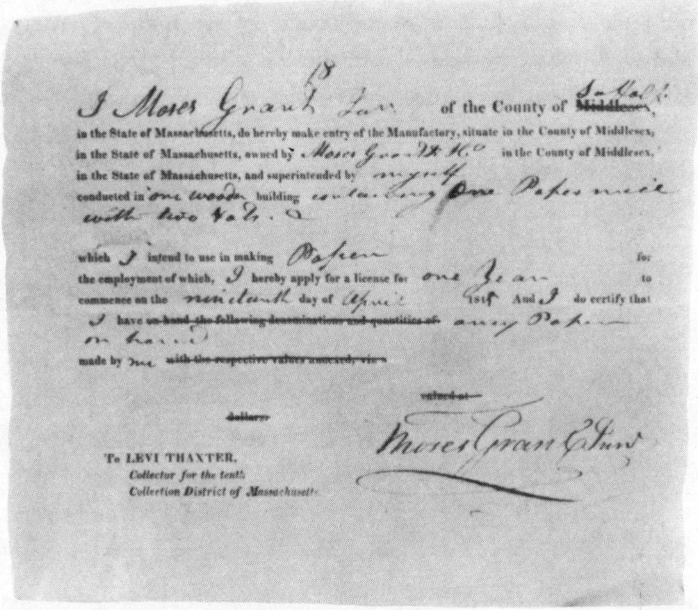

I Moses Grant Jun of the County of ~~Middlesex~~,
in the State of Massachusetts, do hereby make entry of the Manufactory, situate in the County of Middlesex,
in the State of Massachusetts, owned by Moses Grant & Co in the County of Middlesex,
in the State of Massachusetts, and superintended by myself
conducted in one wooden building containing one Paper mill
with two vats.

which I intend to use in making Paper for
the employment of which, I hereby apply for a license for one Year to
commence on the nineteenth day of April 1815 And I do certify that
I have ~~on hand the following denominations and quantities of~~ any Paper
on hand
made by me ~~with the respective values annexed, viz~~

~~valued at~~

~~dollars~~

Moses Grant Jun

To LEVI THAXTER,
Collector for the tenth
Collection District of Massachusetts

NO. 162

An application for a license for the making of paper. The collector will find many items of this kind all of which go to make up an assemblage of "paperiana."

Papermaking in America early grew into a flourishing industry, at the close of the eighteenth century there were numerous small paper mills, and by the year 1810 there were no fewer than 185 different establishments where paper was made.

Today England leads the world in the production of handmade paper, not only in quality, but in quantity, there being seven mills devoted to this craft. Of the 1034 vats that existed in Germany in 1846, but four remain in use, while in Holland, always a great papermaking country, there are only three or four vats in daily operation. Italy has a number of mills, but the paper is not of the best; while in France there are two small mills, and in Sweden but one. In the whole of Europe there are not over 200 workmen who would be capable of forming and couching paper, and the number is diminishing every year. Nevertheless, as long as there is a demand for paper, the handmade product will be needed, as it supplies in the artistic development of the world something which the machine will never be able to duplicate.

PART VI

PART VI

AN ACCOUNT OF THE FIRST ENGLISH FORGERIES AND THE EFFORTS OF SIR WILLIAM CONGREVE TO PREVENT COUNTERFEITING

THE Bank of England was established in 1694, but it was not until sixty-four years later, or in 1758, that the first forgery of its bank notes was practiced. This original attempt at deception was in no way a skilled achievement where the perpetrator was versed in papermaking, watermarking, and engraving, but was the crude attempt of a Stafford linen draper, Richard William Vaughan, to change the figures of a note to a higher denomination. Vaughan was found guilty of the crime and was executed at Tyburn on the 11th of May, 1758.

There were few who wished to follow in the footsteps of Vaughan, so that for twenty years counterfeit bank notes were exceedingly scarce. The next case of counterfeiting bank paper in England occurred in 1778 when John Mathieson, a native of Scotland, tried to manufacture spurious money in duplicate of that used by the Darlington Bank. Mathieson's methods were far more skillful and ingenious than the rough and cumbersome attempts of Vaughan. Mathieson was versed in engraving, and while he did not actually make his own paper, he devised imitation watermarks in plain paper which went undetected for a considerable time, as this mode of issuing fraudulent bank notes was entirely new to the authorities and they could not conceive how their notes could possibly be imitated. In Mathieson's clever work the engraving was finely executed and the watermark, which the bankers had considered an infallible criterion, tallied so precisely with the original that no difference could be perceived. Several papermakers were of the opinion that the devices were genuine watermarks and had been placed in the paper during the process of forming the sheets, but Mathieson declared that this was not the case and that the marks were the result of a particular process known only to himself. After his apprehension he offered to explain the secret of his discovery provided the Corporation would spare his life, but his proposal was rejected and his "secret" died with him.[1]

A watermark may be imitated by copying the required design with a pointed stick, dipped in the following preparations: spermaceti and linseed-oil, equal parts, melted together in a water bath and then stirred until cold; or equal quantities of turpentine and Canada-balsam, well shaken together until dissolved; or the megilp used by artists.[2] If the required designs or symbols be well drawn they will have some resemblance to a genuine watermark as the liquids, when dry, render the paper somewhat transparent. It is, of course, needless to say, how easy of detection all these manoeuvres are, as a mark produced in this way will fade completely when the paper is dipped in water, while an actual watermark will become more brilliant and more discernible. Mathieson, the forger, no doubt employed something of this sort in his work and it is surprising that his simple method of creating artificial watermarks baffled the officials.

In 1773 an act was passed making the penalty death for copying the watermark in bank-note paper; and in order to prevent imitation, it was enacted that no person should prepare any engraved bill or promissory note containing the words "Bank of England," or "Bank post bill," or expressing any sum in white letters on black ground in resemblance of "Bank paper," under penalty of imprisonment for six months.[3]

1 *History of the Bank of England, its Times and Traditions.* John Francis. 2 volumes, London, n. d.

2 *A Guide to the Collector of Historical Documents, Literary Manuscripts and Autograph Letters, etc.* Rev. John Scott and Samuel Davey, F.R.S.L. (page 96), London, 1891.

3 *History of the Bank of England* (page 178).

NO. 163

Charles Price, the forger, in his usual dress and in one of his many disguises. Price was the first counterfeiter in history to attempt the making of paper for false bank notes

Notwithstanding all the precautions taken and the penalties enacted by the British Government after their first experiences with counterfeiters, there was much deceit practiced. It was only necessary for one or two unscrupulous persons to pave the way with their ill deeds and pay the penalty with their lives, for others to wish to do likewise.

The next forger in the annals of the Bank of England was a man of rare skill as a designer, engraver, papermaker, and printer, and while his work was eventually detected he evaded the authorities on every side. Charles Price, or "Old Patch" as he was called on account of a black cloth over one eye, which he wore as a disguise, was one of those men whose whole abilities were employed in defrauding. His first efforts in counterfeiting were practiced about the year 1780 and his false notes were so skillfully engraved and the watermarks so perfectly executed that the Bank of England accepted his bank notes without question, and they were discovered to be forgeries only after they had reached a particular department. Price carried on his unlawful work in Titchfield Street, London, and there had set up a small paper mill where he formed the sheets of paper with the forged watermarks. He was finally taken into custody, and his end was worthy of his life, as he employed his son to procure the necessary implements of destruction and he was found hanging in his cell in Bridewell Prison.[1]

In September, 1801, the following advertisement appeared: "All the one and two pound notes issued by the Bank of England, on and after the first of August, will, to prevent forgeries, be printed on a peculiar and purposely constructed paper; consequently, those dated 31st July, or any subsequent day, will be impressed upon paper manufactured with waved or curved lines."

In the twenty years prior to 1817 there were no fewer than 870 prosecutions connected with banknote forgery in Britain, three hundred persons being executed. In the Bank of England alone there were seventy clerks employed in detecting forged notes, and the year 1818 was the culminating point of the crime. In the first three months of that year there were 128 prosecutions by the Bank, and by the end of the year thirty-two individuals had been hanged

1 *An Account of Forgeries on the Bank of England*. Samuel Thompson. Manchester, 1860.

Charles Price, and his Forgeries on the Bank of England. London, 1786.

for note forgery. From January 1st, 1812, to April 10th, 1818, there were 131,331 pieces of paper forged.[1]

The elaborate watermarking of the genuine bank notes did not seem to deter ambitious counterfeiters, and all sorts of means were employed by them to endeavour to produce something that resembled a watermark.

In 1818 there was great consternation among the Bank of England officials, as well as among the merchants and tradesmen, regarding the extent to which forgers had carried their skill. The engraving and printing did not seem to hinder the efforts of a man who wished to turn his ingenuity to such a purpose; neither did the complications of papermaking and watermarking have a tendency to put a stop to the unlawful practices. The leading typographers and engravers of the day were engaged in trying to produce bank notes that would not easily lend themselves to being imitated, and interest was aroused to such an extent that the Society of Arts attempted to supply remedies to counteract the spread of counterfeiting and published a report in 1819.[2]

No artisan in England laboured harder or more conscientiously than Sir William Congreve to arrive at some peculiar or technical complication in engraving or papermaking that would be a stumbling-block to even the most talented of forgers. While his plans were never adopted by the Bank, they were, nevertheless, of great importance, and it was through his knowledge, patience, and skill, that coloured watermarks were invented and brought into existence at so early a date, for at the beginning of the nineteenth century watermarking in England had not reached any marked degree of perfection.

1 *The Gilbart Prize Essay on the adaptation of recent discoveries and inventions in science and art to the purpose of practical banking.* Granville Sharp. London, 1854 (page 242).

2 *Report of the committee of the Society of Arts, &c. together with the approved communications and evidence upon the same, relative to the mode of Preventing the Forgery of Bank notes.* London, 1819.

NO. 164

Sir William Congreve, the inventor of triple paper. From an engraving by Clint, after J. Lonsdale

Sir William Congreve was born in Staffordshire in 1772. He was a general of artillery in the English army and was attached to the Royal Laboratory at Woolwich. Sir William was a man of remarkable inventive genius and is now best known for his development of the rocket as a military weapon and for his writings relative to warfare. He died in France in 1828.

Congreve's first method of making coloured watermarks consisted in couching a thin layer of white paper and then laying an-

NO. 165

Sir William Congreve's bank note made with three dippings and three couchings. The centre is of Adrianople red enclosed between two thin sheets of white

other tinted sheet in the form of a design, on the first wet sheet; then another white sheet was couched over this which made a triple sheet. After pressing and drying, these layers became a homogeneous piece of paper and the middle coloured layer could only be perceived when the sheet was held to the light. The lettering or device in the second, or coloured sheet, was made by the use of a stencil placed over the mould, cut in the form of the required design. As many colours as desired were inserted between the two outside layers of moist paper, each colour being formed separately.

In 1818, the year in which counterfeiting had reached its height, Sir William tried to have his triple paper adopted by the Bank of England for their bank notes. All of the experimental sheets of paper which he made to show the directors of the Bank, as well as his appeal to that institution to adopt his plan, were set down in a folio volume, the manuscript being in his own hand. This unique book has been in the author's possession for a number of years, and owing to its interesting nature the text is here reproduced verbatim:

"Account of the Origin and Experiments of the Triple Paper invented by Sir William Congreve and proposed for the new Bank note.

"The first suggestion of this paper was given by me to the Commission in a Memoir read to them on the 30th October 1818. In a subsequent memoir dated the 11th December I stated some further particulars respecting this Paper and proposed extracting the colour partially in the interior layer of the Paper by means of acids. Experiments to this effect will be found in the annexed collection. On the 29th December I procured a quantity of Adrianople red cloth, which I was induced to consider the best colour for reasons that will be found to be stated at full length in the following Memoir addressed to the Bank and sent this cloth to Mr. Harman with directions to Mr. Brewer. Mr. Brewer proceeded immediately to Freefolk and made his first report to me on the 7th February 1819, after which a series of letters passed between us containing directions from me to him from time to time, and his observations on sending me back the results as per copies dated February 8th, 10th, 12th, 15th, 16th, 19th, and March 19th.

"With this cloth Mr. Brewer went through a regular series of experiments under my directions. Specimens of which are here annexed with proper explanations. These experiments were at first all carried on by making the triple Paper of three separate layers and three couchings. On the 12th of February I desired to have some with two layers only to be sent up that they might be printed on the interior layer and returned for another white layer, so that the printing would be in the heart of every sheet of paper.

"On the 19th of February it occurred to me that a good effect might be produced by leaving part of the interior bare so as to produce a coloured border round the edge. This border helps to complete the genuineness of the triple Paper with the colour in the interior as is seen in the Memoir to the Bank. In these experiments with the red pulp a very curious effect was accidentally discovered by which the interior coloured layer is mottled.

"In the middle of April Mr. Portal endeavoured to produce an imitation of the triple Paper made on a small scale. A specimen is annexed and it will be seen how unsuccessful an attempt it was. He did not pretend that he was not obliged to go through all the process used by us in making regular triple Paper, or that it did not require Papermakers to fabricate it, for in fact he employed the same men that had carried on my experiments, the only difference was that it was done in smaller quantities and in truth that the effect in making it in smaller quantities is visible enough in the specimen of his imitation. (Messrs. Portal have manufactured the bank note paper for the Bank of England since 1725.)

"This imitation, however, became the subject of discussion at the Commission whose faith in the security derivable from the triple Paper except by the regular process of Papermaking, was rather confirmed than shaken by this attempt of Mr. Portal, and it was subsequent to this that they recommended the adoption of this triple Paper to the Bank.

"But to proceed in detailing the course of the experiments,—all the specimens hitherto produced were either without watermark, or with common wire watermark. I now conceived that a very superior watermark might be produced by a filigree pattern cut out in metal, and accordingly on this principle an oval watermark in silver was prepared by Mr. Branston. The great security of this description of watermark is that the lines of the pattern may cross in the most complicated scroll or cheque work which involves a difficulty in the imitation either by varnish or by cutting out the middle leaf.

"The next improvement which I suggested by letter to Mr.

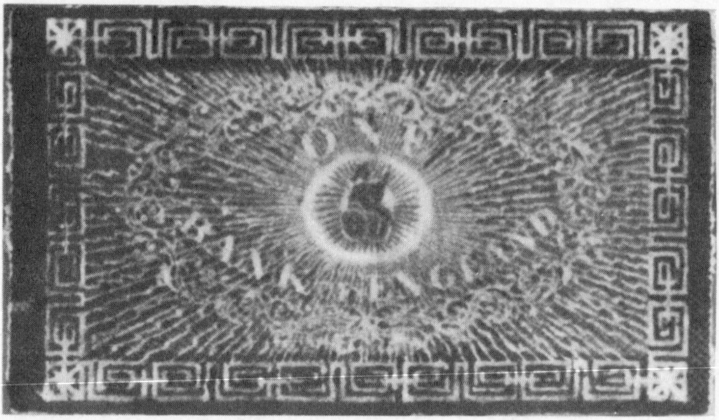

NO. 166

A form of watermarking invented by Sir William Congreve about 1818. The filigree, cut in silver, was removable and acted as a stencil

Portal of the 15th June was to attempt to make the triple Paper by three dippings and only one couching which I conceived would not only improve the brilliancy of the watermark, but would also save labour and expense in making the Paper by saving the operation of *two* couchings out of *three* in each sheet, and I accordingly prepared another watermark for the experiment. The result of this experiment fully justified my supposition and some specimens are attached in which I think the art of Papermaking for Bank Notes is brought to the highest state of perfection.

"I caused another watermark also to be made which was used quite on a different principle to any hitherto adopted, it was not attached to the mould, but to the deckle and lifted off with it, so as to carry away with it all the pulp where the watermark is intended to appear."

There are thirty-six specimens of paper affixed in the back of Congreve's manuscript volume. The first twenty-four examples were fabricated by the original method of three dippings and three couchings. These are the identical sheets which were fashioned by Mr. Brewer for Sir William, and were made, as Congreve states in his explanation, " . . . with common dye; the paper thick and clumsy." Other specimens shown were produced, "with the interior pulp made from Adrianople red cloth and the paper much finer." There are also sheets of two colour combinations, made with white and yellow, and white and red pulp. Congreve shows one sheet in four colours, having used red, yellow, blue and white material. This particular sheet of bank note paper was the most complex ever attempted up to that time, and although Sir William stated that it "is crude and rough," the specimen has many good qualities; and when it is considered that this sheet of paper was made in the infancy of coloured watermarking, it must be regarded as a remarkable specimen. Congreve also gives specimens of paper with printing on the interior of the sheet, giving the appearance of a watermark in black. This was accomplished by printing with common printers' ink upon one leaf and then couching another leaf over the printed one. He also extracted the colour from the ink, leaving the lettering in white, resembling a genuine watermark made with wire. Of the attempt by Mr. Portal to imitate the Congreve triple paper, the inventor has this to say: "Mr. Portal's imitation of the Adrianople red triple Paper, the badness of which imitation will be evident by comparing it with one of the genuine Adrianople notes below it." While Sir William's specimen is a little more brilliant in colour than that of Mr. Portal, I would not suggest that the "imitation" was nearly so bad as Sir William would lead us to believe.

In advocating the use of his triple paper in the making of bank notes, Sir William Congreve directed this letter to the officials of the Bank of England:

NO. 167
Congreve's bank note paper with a pink border and white centre

"TO THE GOVERNOR AND DIRECTORS OF THE BANK OF ENGLAND:—As the triple Paper which I have had the honour to propose and superintend for the new Bank note seems now to be brought very nearly to perfection, I think it is desirable that I should give as concise and summary a view as I can of the principal points of security which I have had in view, in proposing this plan, and as all the points now left for decision rest entirely with the Governor and Directors of the Bank, I have thought it right to address this paper in particular to them.

"First then, as to the security arising from the mode of fabricating this paper. I feel confident that no imitation of this paper with the layer of coloured pulp thrown into the interior, can be made without going through the process of papermaking, and indeed this has never been denied. The imitations that were attempted were made as paper and by papermakers, and no man has ever been bold enough to say he can produce the effects here produced by any process subsequent to the original formation of the paper.

"This fact alone therefore, amounts to no ordinary security for most assuredly the forger cannot as at present by various simple means take a piece of common paper and produce the appearance of a watermark upon it. He must, as I have already observed, absolutely go through the process of papermaking, and moreover to produce the extraordinary clear watermark thus given he must discover and pursue a process quite new and little likely to be suspected even by an expert papermaker, of dipping the three layers of pulp, one upon the other, without couching, and still further of making the coloured layer in clear water. I say, therefore, that as this new and extraordinary system of manipulation has been found essential to the production of the new watermark so peculiarly clear and transparent, there is no probability, for the present at all events, of its being imitated even by a papermaker.

"In the second place, as to the security arising from the introduction of colour. It is evident that the tint in the interior of the paper gives a brilliancy to the watermark which cannot be obliterated by the wearing of the note, or by its being soiled, whereas in the present white note, after being considerably rubbed and soiled, it is extremely difficult to distinguish the watermark. Another very important advantage in the introduction of colour in the interior of the note is that it is a much greater security than the thinness of the present paper against attempting any alteration in the value of a note, such as the making a ten pound out of a one or two pound note by erasure. This is a mode of forgery that has been practiced with the present paper, but with the coloured paper any erasure which would not show on colourless paper would produce a greater strength of colour by laying bare the interior.

"With regard to the particular colour, the pale blue as far as appearance goes seems to be the most preferred, there are, however, reasons which induce me to prefer the pink produced by the Adrianople red dye, and which I shall here state as this is a point resting with the Bank, and one which I think well worthy their mature consideration as independent of the general security attached to the introduction of colour, much of the security depends upon the use of this particular dye. The fact is that the pink pulp with which the first specimens were produced was of a very peculiar and remarkable tint that can only be obtained from the Adrianople red and is moreover a colour that cannot by any possibility be applied to the pulp after it is made, but must actually have been given to the cloth previous to its being made into pulp. The least quantity of pulp that can be made at one time is one hundred weight.

"As therefore to obtain pulp of this colour so large a quantity must be made at once, it is evident that the true Adrianople pink pulp, which can always be distinguished from any other tint, can only be made in a regular paper mill on the largest scale. The Adrianople red cloth from which this pulp must be obtained is not produced in more than two or three principal manufactories in this country, as the process of dyeing it is a most laborious, troublesome, and uncertain operation consisting of nine or ten different manipulations. By adding this dye, therefore, the security arising from the paper appears to be completed, for the forger would not only then be obliged to make the paper, but to make it in large quantities. I must confess, therefore, that I am of opinion that although there seems a predilection for the blue tint, that this is a point which should be reconsidered and especially as the tint may be given in as light a shade as the blue, and as little detrimental to the effect of the printing.

"Test of the genuineness of this Paper:—The most simple rule may be laid down for the test of the genuineness of the triple paper, namely that when held up to the light and looked through the colours will look much stronger than when looked at. Now, if the colour were not in the interior, which we have seen is a process too difficult for the ordinary forger to attempt in his paper, the

NO. 168

Bank note paper fabricated by Sir William Congreve in 1818. This was the first coloured watermark. The layers of paper are blue, red, and yellow, enclosed between two layers of white

very reverse would be the effect, that is, if the colour were stained on the surface which seems to be the only mode of imitation open to the forger, then would the colour look paler when looked through than when looked at, instead of looking stronger as in the genuine note. And to prove this difference to the Public on the note itself, a narrow border of the interior coloured pulp is left bare all round the note, in this border, therefore, the colour is superficial and accordingly when the note is held up to the light the border where the colour is superficial and which is the strongest tint when looked at, appears the palest when looked through, and vice versa, the remainder of the note, where the colour is in the interior and which appears the palest when looked at, is much the strongest tint when looked through. Thus the truth and value of this simple test are at once established on a first inspection.

"Of the expense:—Mr. Brewer has informed me that it has been ascertained by experiments in the presence of Mr. Portal, that one man could on the last new plan of three dippings with only one couching make eight sheets of this new paper in ten minutes without any succession of moulds. The following is the calculation he founds upon this fact: 8 sheets in 10 minutes by one man; 48 sheets in one hour; 480 sheets in one day of ten hours. If there be eight notes on a sheet, 3840 notes may be made in one day by one man and if 20 men are employed they would make 76,800 notes in a day, which is, I believe, considerably more than required."

The Governor and Directors considered seriously the use of Sir William Congreve's invention for their bank notes and conducted a number of experiments regarding its probable use. It appears that Mr. Portal, whose firm had long since made all of the Bank of England paper, did not take kindly to Sir William's new triple paper and dissuaded the Bank officials from its adoption. Producing the triple paper would have involved great difficulties and had the Bank authorities and Mr. Portal seen fit to decide upon the new paper, its fabrication on so large a scale would have been a perplexing task for even the most adept of papermakers.

Sir William Congreve was naturally disappointed when the Bank officials finally refused to make use of his paper for bank notes, and in a communication addressed to the Commission dated September 11th, 1819, he writes:

"The main points as to the fabric of triple paper having been successfully accomplished, that is to say the cleanness and brilliancy of the watermark and the fitness of its texture for a bank note having been brought to perfection, its adoption for that purpose having also been decided upon by the Commission, it was natural to suppose that time would have been given to realise the necessary arrangements for making it in sufficient quantity for the supply of the Bank. No preparations, therefore, or experiments were made for this object which was considered as the minor point until about a fortnight since, nor indeed was any opposition anticipated on this ground since the first objections stated were that the paper was too easily made, and not that it was too difficult. The event, however, has proved there was a want of due precaution and foresight in this want of preparation, for seeing the determined opposition that is made to it in every stage, arrangements for producing the supply ought certainly to have been made and proceeded hand in hand with the main operation of perfecting the paper.

"As the matter now stands this paper has only been produced in a perfect state with the two note moulds. This much, however, has been ascertained from the experiments that have been tried with the eight note mould, that it is the breadth of these moulds and not their length that prevents the man from throwing off the water as well from an eight note, as from a two note mould. Now, if the notes of the eight note mould were put crossways instead of lengthways, the eight note mould might be made a very little broader than the two note mould, and might therefore be used equally well in the production of the triple paper.

"I am, therefore, still convinced that the necessary quantity may thus be made with very little additional increase of means at Freefolk. I have no wish, however, to press the matter further on the Bank and Mr. Portal in defiance of such determined opposition on his part, if the Commission think fit to rescind their decision for its adoption. I have only to say that if such be the result I shall find other means of ascertaining for my own satisfaction after having taken so much trouble how far the fabrication of this paper can be realised on the great scale and at what expense."

Signed, WILLIAM CONGREVE.

As Mr. John Portal figured so prominently in the controversy between the Bank of England officials and Sir William Congreve regarding the triple paper, it is interesting to trace the history of his family and his record and experience as a papermaker.

The Portals were a Huguenot family and were of Albegeois descent, their house, both ancient and noble, having long been associated with Toulouse. During the trouble under Louis the Fourteenth, Louis de Portal, then head of the family, attempted to escape, but he and his wife and one child were killed, and only the four other children managed to flee to Holland. One of these, Henri, attached himself to the Court of William of Orange, and when the latter came over to England to supplant James the Second, he followed after a lapse of several years. Henri landed in Southampton in 1706 and found employment in the paper mill at South Stoneham which belonged to the Governor and Company of White Papermakers in England. Here Henri de Portal mastered the many branches of papermaking. In 1711, when he attained his majority, Henri de Portal was naturalized and became Henry Portal. When he started out for himself, Henry first had the Bare Mill, near Whitchurch, but in 1718 he took over the neighbouring Laverstoke mill, which the family has held ever since. The original tenure was a ninety-nine years' lease at £5 and "one Reame of Fools-cap paper, neatly cut" per annum. Under the lease he was required to rebuild the mill, which he did in 1719. In 1725 Henry Portal secured the monopoly for making bank note paper for the Bank of England.

The bank notes that were manufactured in England previous to Henry Portal's contract were of plain paper, without watermarking. Those from the Laverstoke mill had a watermark border of a loop pattern running around the edges of the sheets, and ever since 1725 the Bank of England notes have been watermarked.

Henry Portal died at Freefolk Pryors, adjoining Laverstoke

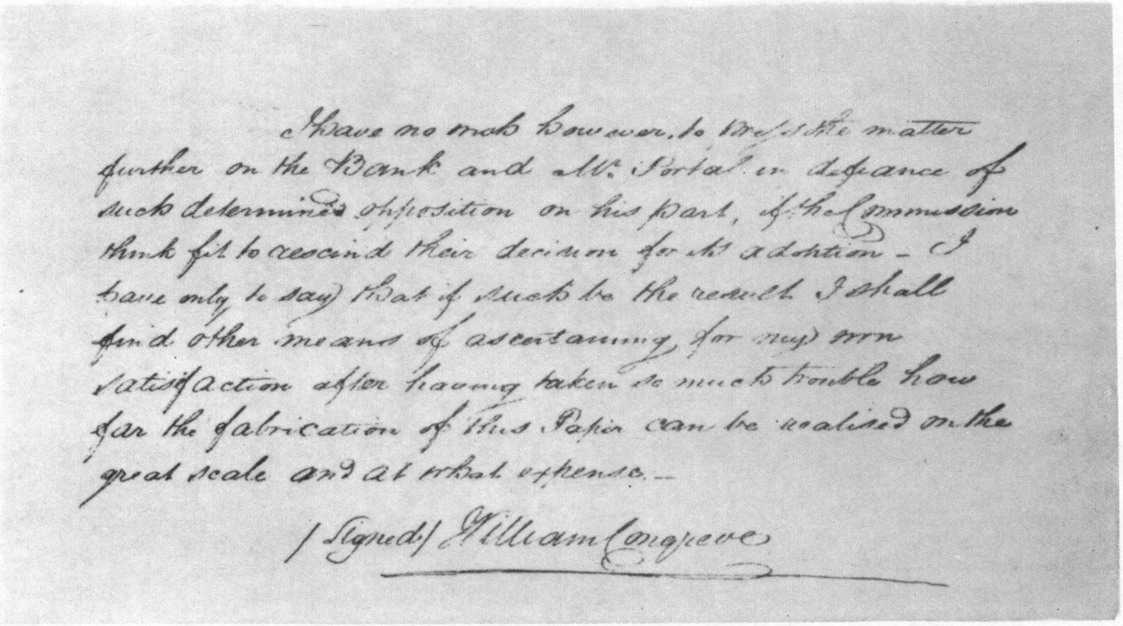

I have no wish however, to press the matter further on the Bank and Mr. Portal in defiance of such determined opposition on his part, if the Commission think fit to rescind their decision for its adoption – I have only to say that if such be the result I shall find other means of ascertaining, for my own satisfaction after having taken so much trouble how far the fabrication of this Paper can be realised on the great scale and at what expense.–

(Signed) William Congreve

NO. 169

The last paragraph of the manuscript addressed to the Governor and Directors of the Bank of England by Sir William Congreve, September 11th, 1819

mill, in 1747. He was succeeded by his son, Joseph Portal, who operated the mill until his death in 1793. His son, John Portal, continued the fabrication of the Bank of England notes for a period of 53 years, until his death in 1848. The paper mill is still continued by the Portals, a record of over 200 years in the same family. It will be seen that John Portal, Sir William Congreve's adversary, had come from a family of papermakers and that his own long experience qualified him to advise the Bank of England authorities regarding the adoption of Congreve's triple paper.

Sir William Congreve was 46 years of age when he discovered that three separate sheets of paper could be formed and couched as one sheet, thus introducing colour into the interior of the paper. There is no record of Sir William having made any other triple paper than that shown to the Governor and Directors of the Bank of England in 1818, all of which is in the author's collection of paperiana. Congreve patented his invention December 4th, 1819, (NO. 4419).

There was but little accomplished in the art of coloured watermarking from the time of Congreve until about 1885 when Mr. Lee of Wookey Hole, near Wells, Somerset, England, undertook to revive and improve the art. Clayton Beadle in his article on watermarking[1] gives Lee the credit for being the pioneer in England in the making of coloured watermarks, but this is obviously a mistake. Mr. Beadle also states that Lee patented this invention in 1886, but a search through the British Patent Office records does not reveal a coloured watermark patent under his name.

As late as 1900 W. Fairweather and A. and G. B. Fornari were granted a patent in England which reads in part: "A sheet of white or coloured pulp, having any design suitable for a watermark, is inserted between two plain sheets of pulp, and the three layers are pressed together to form a single sheet of paper. The

1 *Journal of the Society of Arts.* London, May 18, 1906.

mould consists of a base over which is stretched the sieve, the latter being covered with a plain plate having the required design cut in it. The watermark can only be seen by transmitted light."

PART VII

PART VII

THE USE AND SIGNIFICANCE OF THE ANCIENT WATERMARKS IN EUROPEAN AND AMERICAN BOOK AND MANUSCRIPT PAPERS

THE study of watermarks attracted little attention until the early eighteenth century when several writers on typographical subjects devoted some research to papermaking archaeology. During the nineteenth century a number of works in French, German, Italian, Russian, Dutch, and English were compiled dealing with the historical importance of the old watermarks, and the last decade has seen an increasing interest in their study. So difficult is the study of old watermarks that, though several antiquaries devoted their lives to the subject, no writer has ever entirely fathomed the mysteries or complexities of these ancient devices. Almost every scholar who has written on the subject has advanced a different theory of their origin, utility, and meaning. Their symbolic aspect has been dealt with thoroughly, and numerous compilers have reproduced and indexed for us thousands of the antique papermarks, but without any definite theory as to their inception and use.

It has been suggested that these old devices may have been used solely as marks of identification for sizes of moulds and the paper formed thereon, or as trade marks of the papermakers who fashioned them. Others have advanced the theory that they may have been employed in a purely symbolic sense, as Harold Bayley[1] sets forth in his books dealing with the semeiotic significance of the old papermakers' and printers' marks and emblems. According to Mr. Bayley the watermarks of the Middle Ages were employed by

1 *The Tragedy of Sir Francis Bacon*. London, 1902.
A New Light on the Renaissance. London, 1909.
The Lost Language of Symbolism. London, 1912. (2 vols.)

the heretical papermakers as symbols of religious propaganda. Europe at that time was overrun with mystic and puritanical sects, and the art of papermaking was one of the most developed trades of these religious bodies. Mr. Bayley says: " . . . The early papermaking districts were precisely those that were strongholds of the heretical sects known as the Albigenses. The word 'Albigenses' is a term applied loosely to the various pre-Reformation reformers whose strongholds stretched from Northern Spain across the southern provinces of France to Lombardy and Tuscany. In Spain and France they were known as Albigenses from Albi, the name of one of the prominent towns. In the Alpine provinces they were called Waldenses, from Peter Waldo, one of their most conspicuous members. In Italy, history alludes to them under the terms Cathari or Patarini." Mr. Bayley attaches symbolic importance to each of these watermarks used by these mystic people and believes that the papermarks carried with them signals of hidden meaning. This seems really more probable than to try to account for the multitude of watermarked designs as marks of identification for paper sizes or as trademarks of the makers.

It is not entirely out of the way to suggest that the old watermarks were perhaps nothing more than a mere fancy with the papermakers, who may have formed the designs or emblems to satisfy their own artistic natures. In the whole craft of papermaking there is no part more interesting or fascinating than to couch a sheet of paper upon the felting and watch the impressed mark become clear and distinct as the water slowly evaporates. Another supposition regarding the use of the early papermarks is that since the workmen could not read it was necessary to appeal to them by means of pictures. Simply to have marked a mould with letters or figures would have meant little to the artisans of the fourteenth and fifteenth centuries; it was essential to convey the meaning to them by the aid of illustrations. For the same reason the old sign-boards of inns and shops were always of a pictorial nature; the name of the tavern lettered upon the swinging sign, or of the commodity sold by the tradesman, would have been of slight effect.

At the present time watermarks are trademarks of the paper-

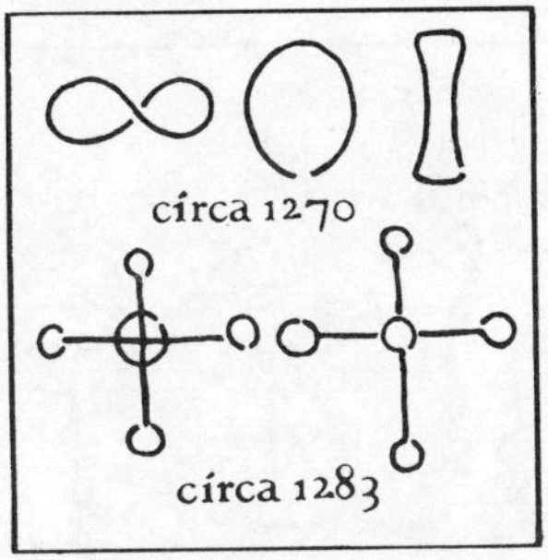

NO. 170

The first papermarks appeared in Italy the latter part of the thirteenth century

makers, pure and simple. It would seem that their ancient significance must remain more or less obscure. It is, however, possible to determine almost exactly the date at which watermarks were first introduced into sheets of paper.

In the ancient Oriental papers, both Chinese and Persian, no watermarks are found. The earliest moulds upon which sheets of paper were formed were constructed of bamboo, a material which did not lend itself readily to making the applied designs which would watermark the paper. This fact may perhaps be sufficient to account for the absence of such marks in Oriental papers. It would appear that in Europe iron wire was used as a mould covering when the art was introduced into Spain, yet papermarks did not make their appearance until more than a century later.

The first papermarks appeared in Italy about the year 1270, and while most insignificant in design the emblems chosen lead us to believe that they were meant for use as some sort of signal or symbol to convey a meaning among the workers who made them, or among those who used the paper in which they appeared. However that may be, the use soon became general, hundreds of different emblems being employed by every papermaking country of

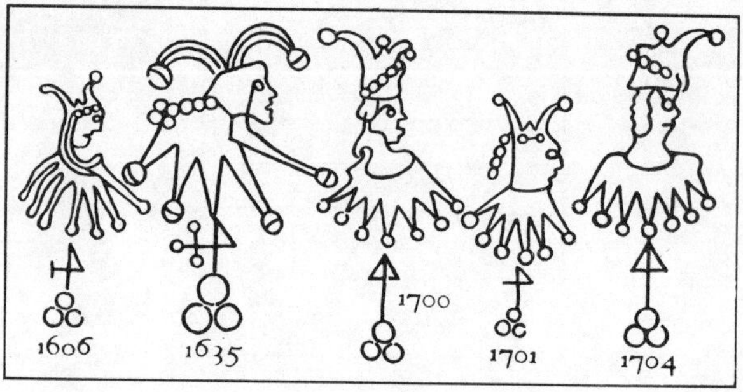

NO. 171

The watermark of the foolscap first appeared about 1479 and there are many versions as to its origin

Europe. These designs multiplied in numbers as the centuries passed, until there were used in paper literally thousands of different ciphers typifying almost everything in nature and every human endeavour. When about 1450 printing from movable types was first introduced into Europe the art of watermarking was an accomplishment almost two centuries old. But when we think of the instances of printing and watermarking in nature from the beginning of time, it seems strange that man did not put these arts of impression to use much earlier,—for did not animals, no matter how remote, leave impressions or stamps of their feet in the sands, and vegetation impress itself in the clay? These things in nature are really very closely akin to the arts of watermarking and printing.

In examining old books one often notices in a single volume a great profusion of watermarks, some fifteenth century works contain no fewer than fifty papermarks in a single book. To account for this some writers contend that it was necessary for the early printers to purchase their paper from various mills in order to get sufficient material for their use, claiming that the mills were small

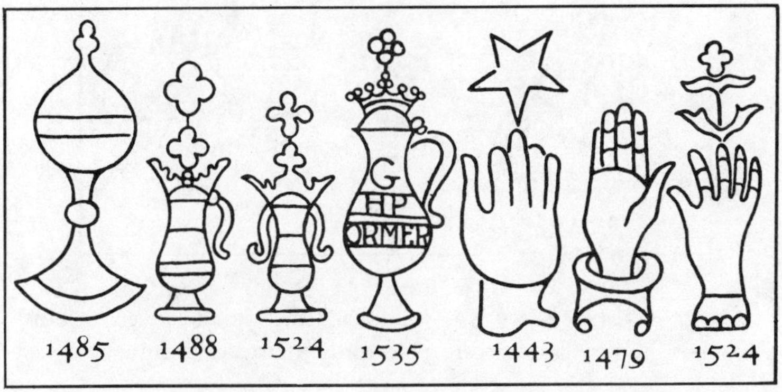

NO. 172

Watermarks of potts and hands were common through several centuries of papermaking

and could not furnish enough paper which was uniform in size, thickness, and watermark. While in the fifteenth century this might have been essential it would not have seemed necessary in the sixteenth; for by the sixteenth century paper manufacture was an industry of vigorous growth in most European countries. It is recorded that John Spilman, who operated a paper mill in England in 1588, employed as many as 600 workers, and England at that time was not a great papermaking country. This may be erroneous for there is not a mill in all England today making handmade paper in which so many employees are needed, and the methods of work do not differ greatly from those of several centuries ago. It is certain, however, that in the sixteenth century supplies of book paper were abundant, yet in single books printed in this century vast varieties of papermarks are found. It does not then seem likely that the marks merely indicated the makers.

It has been suggested that the different watermarks denoted various sizes of paper. This does not seem reasonable, as for every

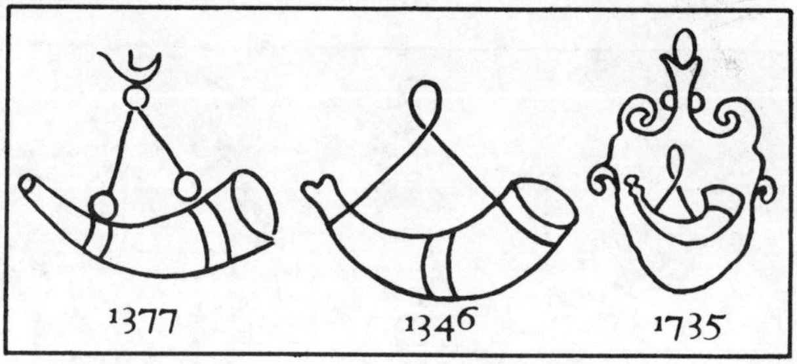

NO. 173
The general post-office was established in England in 1670, but the post-horn watermark dates from the fourteenth century

size of paper a special pair of moulds was needed and the labour and expense of fashioning these moulds were too great to permit such extraordinary variety of dimensions. In later years, to be sure, such marks as foolscap, hand, post, pott, etc., did denote sizes of paper, but these appellations were not used as size names until many years after the establishment of these particular emblems as watermarks, moreover there were many mills using the same designs on different sizes of sheets. The foolscap mark traces to 1479, but in England was later replaced by the figure of Britannia, the mark for this size of paper. Watermarks of potts or jugs are seldom found after the seventeenth century, being replaced by the Dutch or English arms. Hands were used extensively by the old papermakers in Germany and the Netherlands, and at times this mark resembled an iron gauntlet or glove, the initials or name of the maker often appearing on the wrist. The post-horn watermark, according to some writers, first made its appearance in the year

1670, when the general post-office was established in England. This is a mistake as marks of this type have been found in documents bearing dates as early as the late fourteenth century.

The art of forming the wire watermarks that were applied to the moulds has been modified but little since the origin of the craft in the late thirteenth century. In Europe, before the eighteenth century, all paper was made on laid moulds, and the sheets so moulded retained the impressions of the laid and chain wires used in the construction of the moulds. Any wire-work, in the form of objects, added to the top surface of this laid and chain wire covering also made impressions in the paper. Why these indentations were called watermarks is not known, as the mark in paper is no more made by the use of water than is the sheet itself. In German the design impressed in the paper is called *wasserzeichen,* which, like the English term watermark, is confusing. In the French language the term is *filigrane,* and in Dutch *papiermerken.* These two appellations are more suitable. Lewis,[1] writing in 1737, always refers to watermarks as "papermarks," and all contemporary German writers called them *papierzeichen.* In the English patent granted to John Phipps, dated 1790, the name "watermark" appeared. This was, no doubt, the first use of this term, and as the name *wasserzeichen* was not used by German writers until about 1810, we are led to believe that the term, watermark, faulty in its meaning as it is, had its origin in the English language.

The wire forms for producing the watermarks, or papermarks, were for centuries held in place on the surface of the moulds by means of threadlike wires stitched back and forth, binding the mark to the laid and chain wires. In much of the old paper it is possible to detect the sewing-on wires around the watermarks when the sheets are held to a bright light. In many of the early papermarks the sewing-wires are pronounced, owing to wire having

1 *Life of Mayster Wyllyam Caxton.* London, 1737.

been used that was almost as heavy as the wire of the mark itself. At least one writer has stated that the wire designs were fastened to the mould surfaces by the use of soft solder, but this appears to be a mistake, as this method of securing the watermarking wires in place was not used until the first part of the nineteenth century. In watermarks from the thirteenth century the simplicity is striking, as at that time the devices were made of clumsy wire which would not admit of much twisting into complicated shapes. During the fourteenth and fifteenth centuries the wires gradually became finer and the designs more detailed.

The value of watermarks as a means of determining the dates of paper, books, and prints, or the locality where the paper was made, is to be questioned. Few of the early watermarks bear dates, and even when they do, the date of the mark must not be accepted as the time of the printing on the paper. The sheets might have been dated in the watermark and then remained in the mill a considerable time before the paper was sold, and after being sold the paper might have been held for years in the warehouse of the printer before being used. The longer that paper made from fine material is seasoned, the better it will take an ink impression, and the early printers were no doubt aware of this peculiarity. Also a dated mould might have been used for many years with the same date, the papermaker not troubling to change the figures. An illustration of this in modern times is seen in a letter to the writer from the late Joseph Willcox, a direct descendant of the founder of Ivy Mills, the third paper mill to be established in Pennsylvania. Joseph Willcox took over the management of the mill in Chester in 1859, and there made the last handmade bank note paper that was produced in the United States. His letter reads in part: "We had an order for an unusual size of paper and the only moulds we had of the particular size were dated 1810 in the watermarks. I did not like to take off that old date so I made a lot of paper with the mark 1810 in every sheet." This made a difference of almost fifty years from the date in the paper and the time when the paper was actually made.

It would also be unwise to rely on a watermark as proof of where the paper was made or at what particular mill. Suppose that an early papermaker established a reputation for a superior quality of paper, there was nothing to prevent a newly set up mill from using the watermark of the older and more prosperous concern. A good illustration of this in our own time is the imitation of the watermark of the highly-esteemed Whatman mill, a concern that was established by James Whatman in Maidstone, England, in 1731. Certain continental papermakers have for years duplicated the Whatman watermark and sold their papers without hesitation, the sheets often being accepted as genuine Whatman papers. It is not unlikely also that in the early years of papermaking moulds were sold by one mill to another without troubling to remove the wire watermarks; or the old worn moulds of a large mill might have fallen into the hands of a less prosperous maker who fabricated an inferior quality of paper. In this way one watermark design might have been used by numerous mills, over periods of many years, for papermaking moulds have always been well and strongly made and they do not easily wear out if treated with any degree of care. From an archaeological and artistic standpoint watermarks are of great interest, but for tracing definitely the dates of paper, or the exact localities of certain mills, the marks must not be relied upon too implicitly.

Almost all writers touching upon the subject of ancient paper and watermarks attach importance to small changes that often appear in papermarks depicting the same subject. This seems to be a mistake, for the slight variations may be due to very simple causes. The wire forms may have become detached from the moulds and have been replaced by a worker unskilled in wire working,—possibly the vatman or coucher. Every time a wire mark came loose from a mould it had to be attached again, and in so doing some detail naturally was slightly changed from the original outline. This alone would account for the large number of

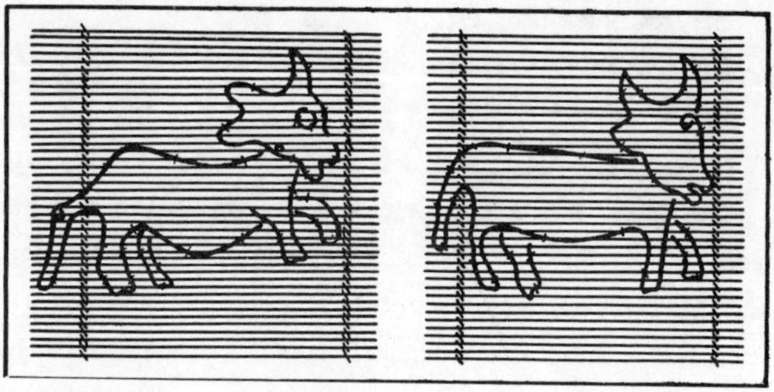

NO. 174

Slight variations in old watermarks should not be regarded as important as some writers lead us to believe

marks, similar in subject, and supposedly from the same mill, but varying to a slight degree. Another simple cause for variants may be suggested. The two watermarks of the bull or ox shown as an illustration, were traced from sheets of paper from a copy of the first volume of the Mazarin Bible, attributed to Johann Gutenberg, and printed between 1450 and 1455. In these two papermarks there is a striking resemblance, no doubt one was a copy of the other, and the two were supposed to be identical. It is probable that so late as the fifteenth century two moulds were used, as at present, in moulding paper at the vat. Each mould had a wire watermark of a bull, the two possibly formed by different workers, each one trying to rival the other in design, but keeping the two emblems within the same limits of space and general contour. The small impressions that appear crosswise along the lines of these devices were caused by the sewing wires which bound the larger watermarking wires to the moulds. These small lacing wires are discernible in almost all old watermarks if the sheets are held to a brilliant light.

The paper that was used in the 42-line Bible, from which the

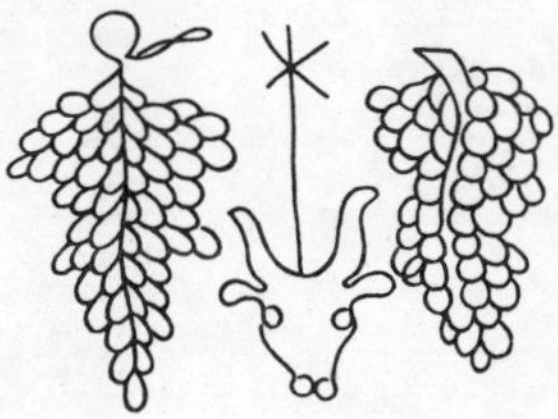

NO. 175
The most prominent watermarks of the Gutenberg 42-line Bible

two bull watermarks were traced, is of the finest quality and in many ways its excellence has never been surpassed. The watermark of the bunch of grapes which is found in much of the Gutenberg Bible paper is unusual in its brilliance and clearness, although the paper is long-fibred, which, as a rule, does not produce sharp and distinct watermarks. The strength of the paper must always be sacrificed if a well-defined and distinct watermark is desired. While the bunch of grapes as a papermark is of Swiss origin, I am inclined to believe that the paper used in the first printed Bible had its origin in Gutenberg's native land. No matter in what locality it was fabricated, or in what particular mill, this paper shows technical skill and workmanship that is seldom encountered in modern times; the texture, strength and colour have remained unchanged over the period of almost five hundred years. I am afraid that through the use of chemicals, much of the handmade paper of our time, even with the 479 year handicap, will suffer by comparison with that of the Gutenberg Bible in another such period, or in the year 2408.

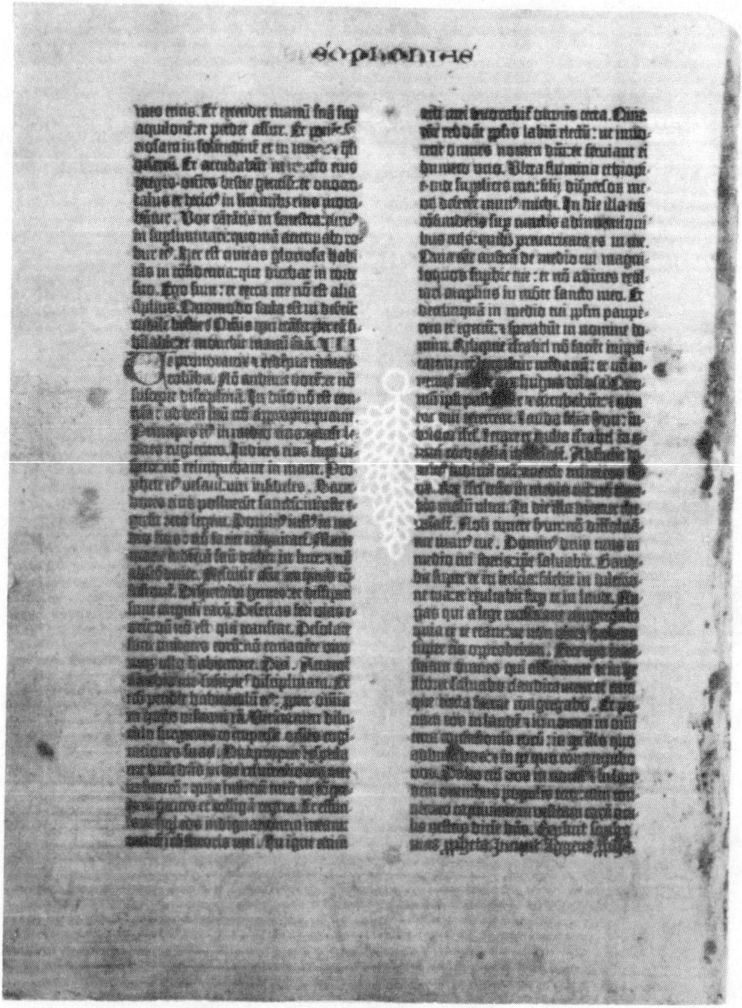

NO. 176

The bunch of grapes watermark in a page from the 42-line Bible printed by Johann Gutenberg, 1450–1455

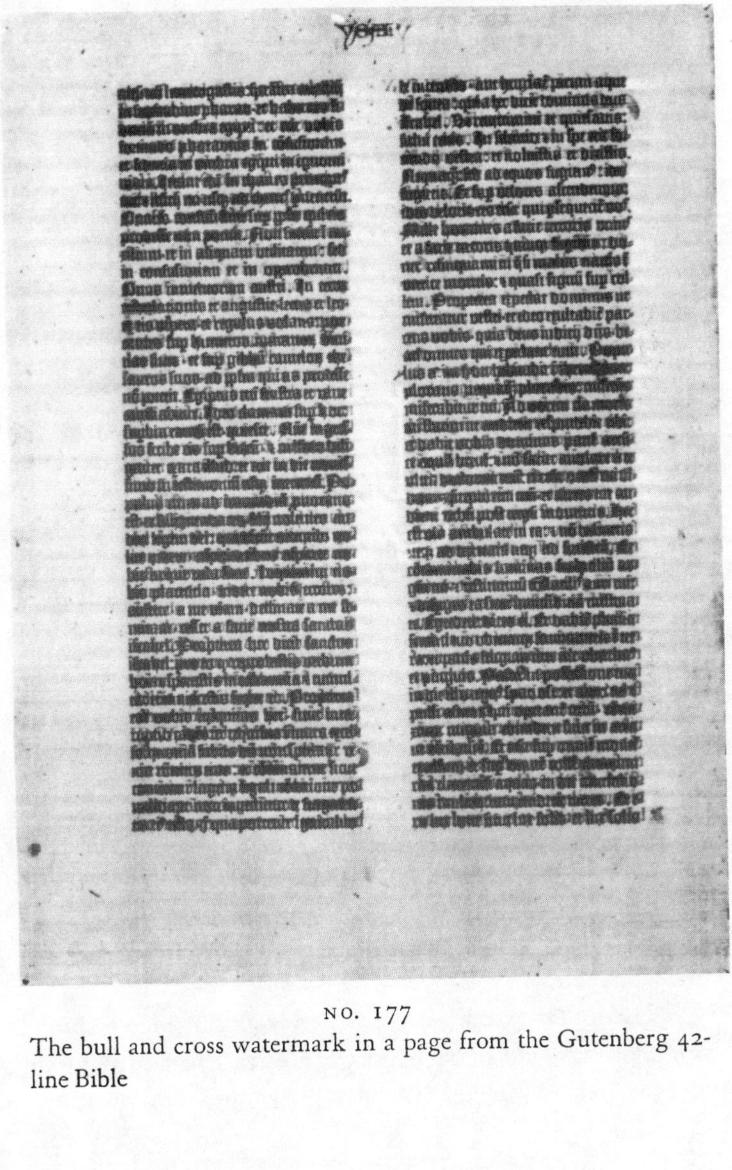

NO. 177

The bull and cross watermark in a page from the Gutenberg 42-line Bible

After watermarking became general, during the fifteenth century, it is seldom that a sheet of paper is found without a distinguishing device. These emblems in most cases were placed in the centre of the sheets, or where the paper was folded in book printing, but it is not unusual to find two symbols or designs each appearing in the middle of the half-sheet of paper. The watermarks, from the origin of the art until the latter part of the eighteenth century when they began to lose their simplicity, may be consistently divided into four divisions. The first of these classes would embrace the earliest known watermarks, which appeared in the form of crosses, ovals, circles, knots, triangles, three-hill symbols, and devices of the simplest kind that could have been readily twisted in wire. At this period there were also a great many Pommé crosses used as emblems for watermarking paper. This was a Greek cross with balls or circles placed at each end of the cross bars. Another similar mark which is found in fourteenth century Italian paper, is a circle surmounted by a Patriarchal or Papal cross. The papermarks of this first group were in use from the origin of the art, about the year 1280, until the first quarter of the fifteenth century.

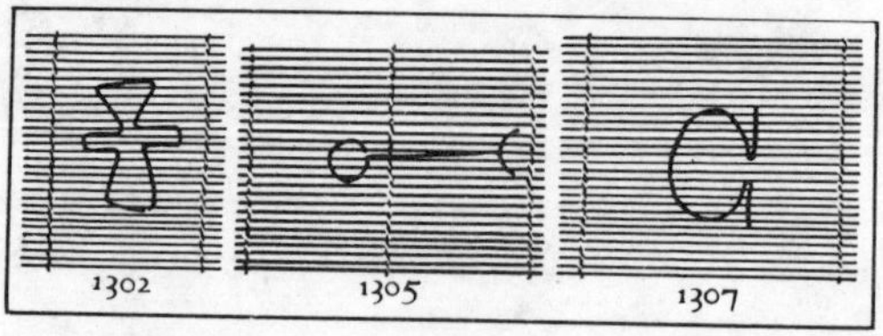

NO. 178

The early watermarks were simple forms that could be readily twisted in wire

The second division of papermarks would include man and the works of man and in the latter class of this particular group we find the greatest number of subjects, extending into thousands upon

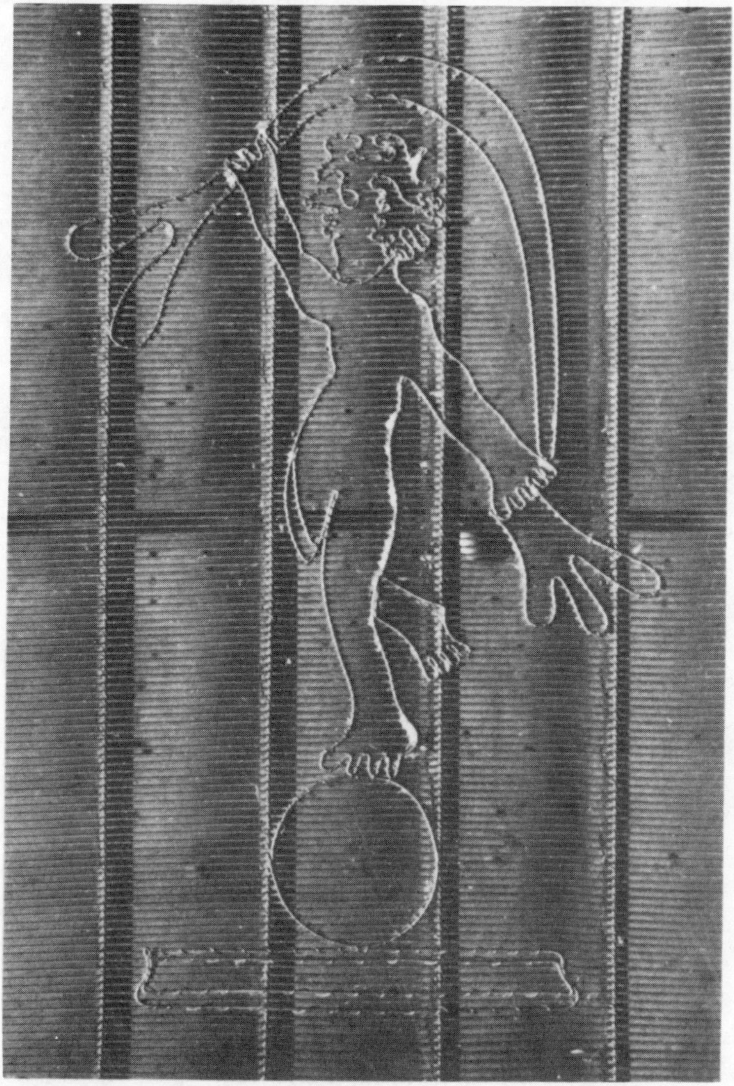

NO. 179
A Dutch watermark of the late eighteenth century. The actual wire on the surface of the mould

thousands of designs. The male figure is met with only in limited numbers and the female figure is rarely found except in mermaid form usually holding a mirror. The human head, feet, and hands were also used separately as watermarks by the early workers. The

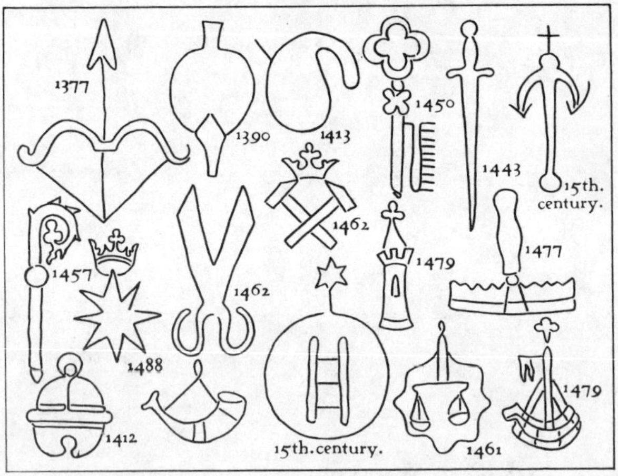

NO. 180

The watermarks representing the works of man would embrace thousands of specimens and a complete cataloguing of marks of this kind would be impossible

mark of the hand was used by the papermakers of various countries for hundreds of years and was symbolic of both Fidelity and Labour. At times the hand shows two fingers bent downward, a sign of benediction. The hand watermark is also found surmounted by a cross, a star, a rose of bliss, or some like ornament, each of which had its symbolic significance. The size of paper known in later years as "hand" derived its name from the emblem, rather than this appellation having been given to a particular size of paper. The watermark of the human foot is exceptional and has only been encountered in a few instances. The works of man

NO. 181

The letter "P" is common in the paper of fifteenth century books and was the initial of Philip the Good or Philip the Hardy. This example is in the paper of *Jacobus Magni: Sophologium* (Hain 10471–10472), printed by Adolf Rusch at Strassburg in 1473

would embrace agricultural implements, and small tools such as shears, spades, bellows, swords, scythes, hammers, pruning-hooks, and axes; with the works of man would also be included such watermarked objects as ships, anchors, anvils, bagpipes, keys, horns, scales, Bishops' and Cardinals' hats and staffs, curry-combs, weapons of all kinds, hawks' bells, as well as architectural ornaments, lettering, and escutcheons. The reproductions given show only a small portion of the multitude of designs that would be classified in the category of the works of man.

In the illustrations showing watermarks of the full figure it has

NO. 182

These watermarks of the human figure were selected from a collection numbering several hundred as the most artistic and representative specimens of the different periods

been the desire to reproduce the most artistic and representative specimens of the different periods, these marks having been chosen from many hundreds of examples. The figures from 1413 and 1426 are of either Dutch or French origin and each typifies an angel, carrying a Maltese cross. The mark from 1457 shows a haloed figure holding a ball surmounted by a Latin cross and is probably German, while the Saint Peter with key, from 1461, might be of either Dutch or French origin. The French papermark of 1477 represents the Virgin holding the child Jesus, but the crudity of execution makes the mark almost undecipherable. The figure from 1500 is meant to portray a pilgrim, having a halo and holding a Latin cross. The next two specimens (1606–1635) are angels, one

NO. 183

A German figure watermark from the middle eighteenth century

supporting a shield with the initials A.P. and an object resembling an anchor cross. The last two marks, both from 1686, are of German origin and were traced from the same sheet of paper.

The earliest watermark of the human head was a portrait of Jesus Christ, and was of French origin, from about the year 1399. This device represents the Vera Icon, or True Image. The legend runs that the Saviour on the way to Calvary was encountered by the woman Berenice. Filled with compassion she wiped His face with her handkerchief, which miraculously retained an imprint of the Divine Features, whereupon Berenice was rechristened Saint Veronica, an anagram of Vera Icon. It is, no doubt, the kerchief with the impress of this head that is reproduced in the late fourteenth century watermark.

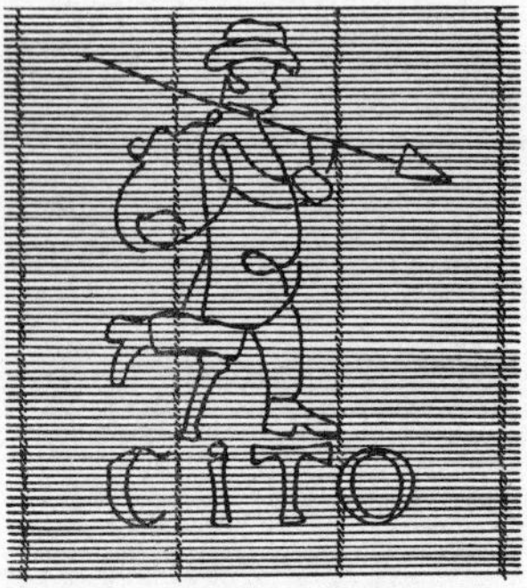

NO. 184
A German watermark of the human figure from 1788

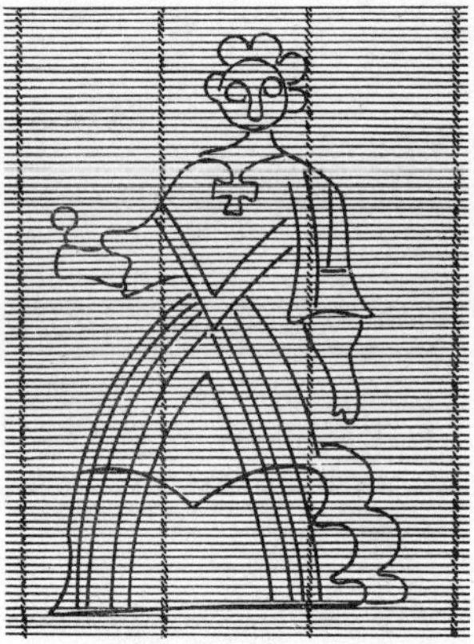

NO. 185
A human figure watermark from the year 1789

Watermarks of Jesus in profile appeared in limited numbers during the fourteenth, fifteenth, and sixteenth centuries. In these marks the head is usually represented with three locks or strands of hair, evidently meant to symbolize Christ's oneness with the Trinity. In most of these marks the mouth is open, intended to denote Jesus teaching. Another profile watermark used in the fifteenth century was that of a negro slave. The specimen shown is from

NO. 186

The original of this figure watermark appears in a vivid green paper fabricated about 1793

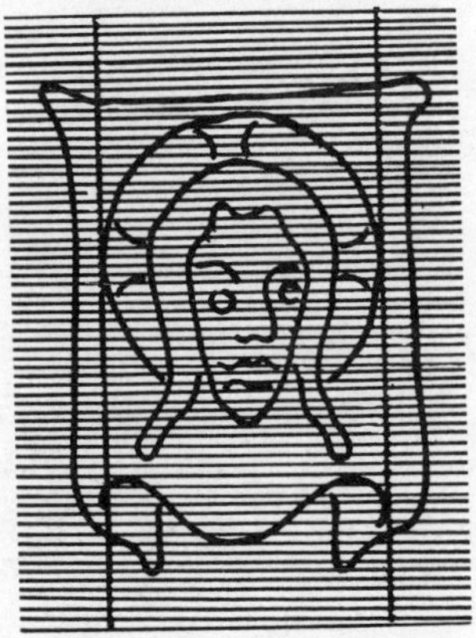

NO. 187

The watermark of the head of Jesus imprinted upon the handkerchief. A French mark of 1399

about the year 1415 and it will be seen that a bandage has been pushed from his eyes, probably intended to symbolize freedom; the head is also crowned, the crown being surmounted by the rose of bliss so much cherished by the old artisans.

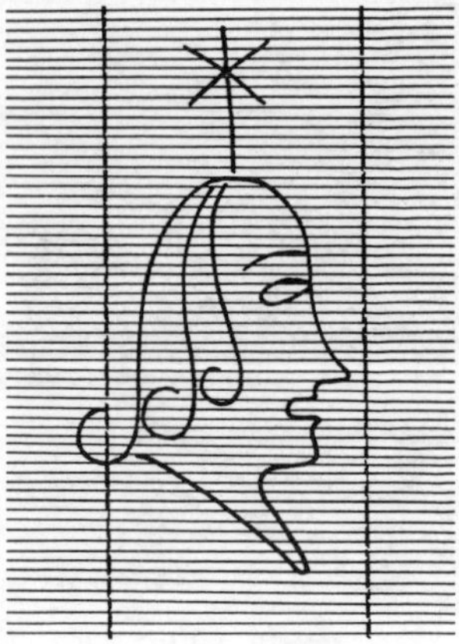

NO. 188

A fifteenth century watermark of Jesus, showing the cross, three strands of hair, and open mouth

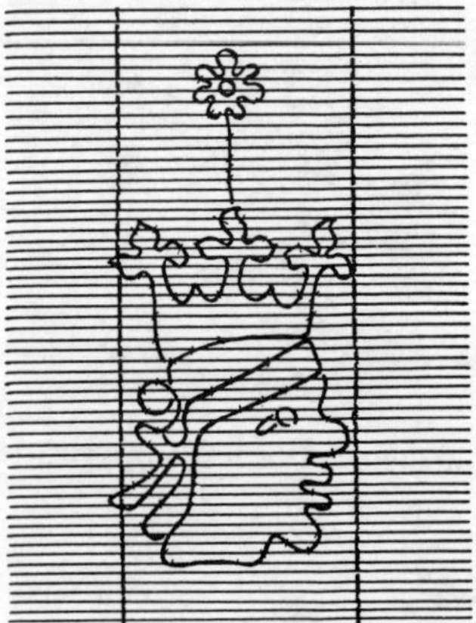

NO. 189

The slave head watermark with bandage lifted. From a specimen of the year 1415

Most of the early watermarks portraying the human figure and head were ecclesiastical in nature, which was natural, as the church penetrated the workshops of the ancient craftsmen in a manner that cannot be conceived today. Nothing really artistic in the

NO. 190

A watermark portrait of King Friedrich, 1750. A representative example of the period

watermarking of portraits was accomplished until the middle of the eighteenth century, when we find the French and German papermakers making outline portraits of prominent personages. The mark of King Friedrich is a representative example of the art as it was practiced as late as 1750.

In arranging papermarks in classes according to their subjects, the third group or division would embrace such marks as flowers, trees, leaves, vegetables, grain, plants, and fruit. In a number of the old emblems there are also combinations of vegetation and the works of man, such as a bunch of grapes with a bell or crown, or a pott or jug holding flowers or leaves. In the reproductions of watermarks of nature subjects the first two specimens, from 1394, are roses, a form of watermark found in paper from the fourteenth to the eighteenth century. The mark from 1360 might be either a melon or leaf, while that of 1390 appears to be a pomegranate. The cluster of berries is from 1435, and the device of the pomegranate between two leaves is from about 1443. The watermark of the

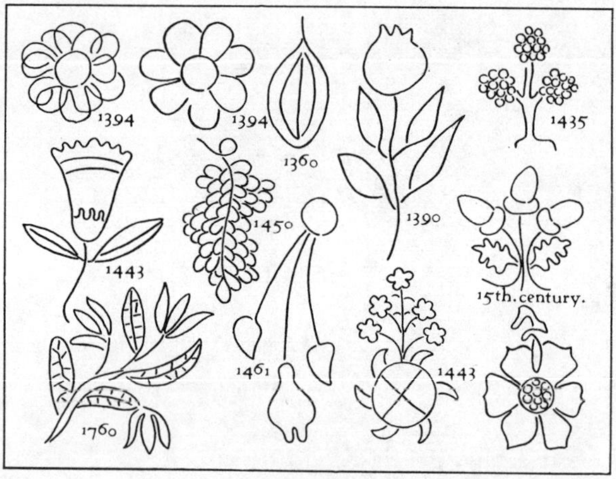

NO. 191

Every form and combination of vegetation has been used by the old papermakers in watermarking their paper

grapes, of Swiss origin, is found in books printed before 1500. The pear between two leaves (1461) was a common papermark with the old papermakers in the fifteenth and sixteenth centuries and occurs in many shapes and sizes, at times very well executed. The specimen of 1443 is the forget-me-not flower over a Catharine Wheel, by the great teeth of which Saint Catharine of Alexandria was tortured A.D. 307. A German writer on the subject of watermarks terms this device a waterwheel, confusing the teeth of the wheel of martyrdom with the buckets of a wheel used for power. The last two papermarks illustrated in this group are from the fifteenth century and reveal considerable ability in both their

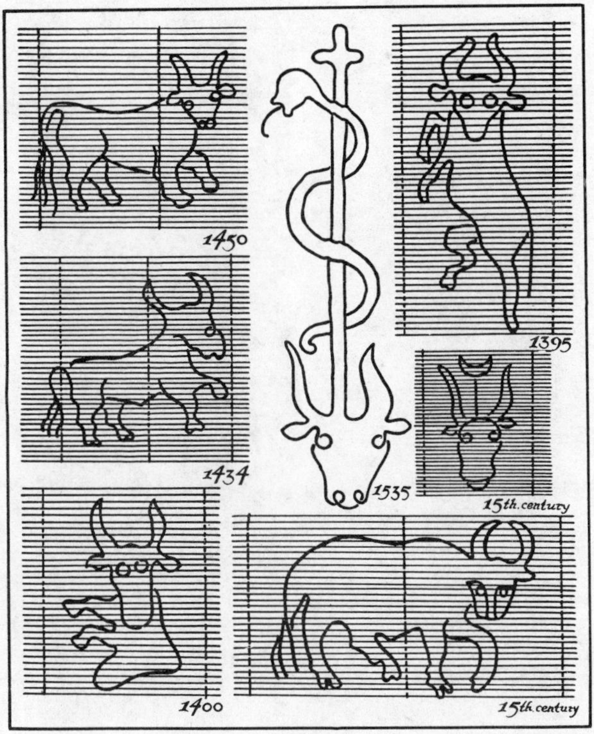

NO. 192

The most interesting group of watermarks would include wild, domesticated, and legendary animals, as well as snakes, fish, snails, crabs, scorpions, and all varieties of insects

drawing and the forming in wire. The nature papermarks became fewer toward the close of the eighteenth century, but this period is remarkable for some very large and complicated floral emblems. The mark of 1760 is representative of this time. These watermarks of natural subjects were selected from a collection numbering several thousand, as best showing the use of devices of this variety.

The fourth and probably the most interesting group of watermarks would include wild, domesticated, and legendary animals,

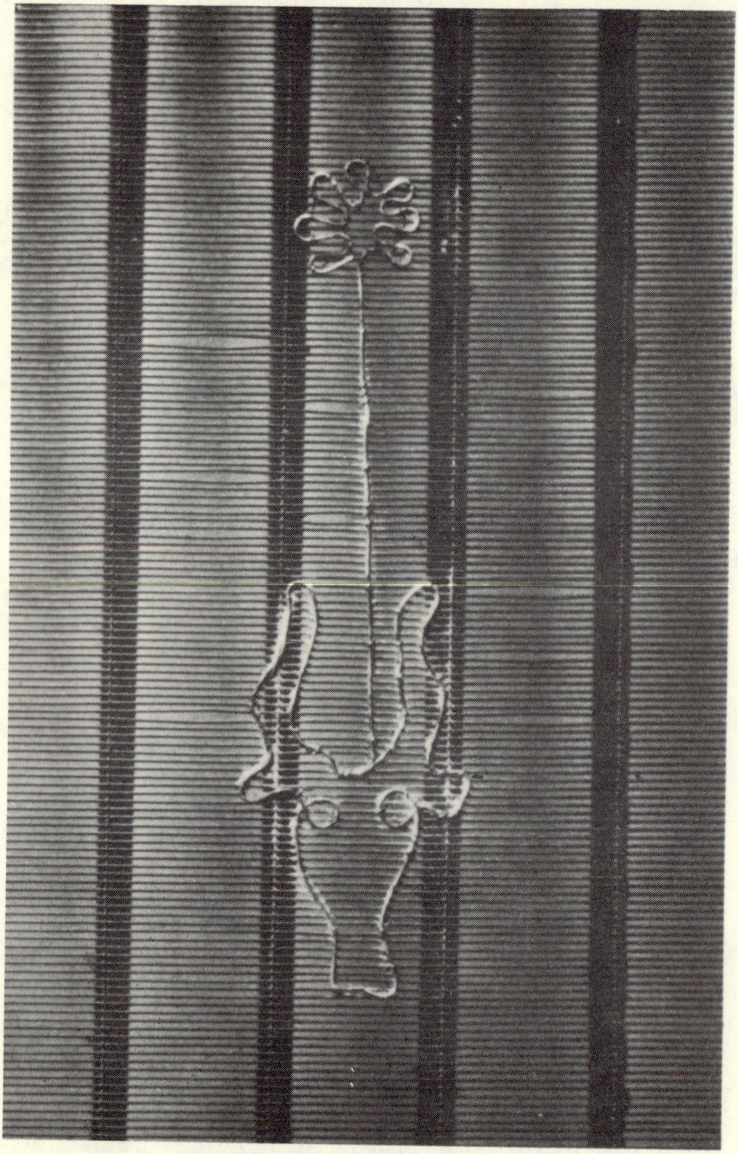

NO. 193

The bull's head surmounted by a rose of bliss. The actual wire form laced to the surface of an antique laid mould

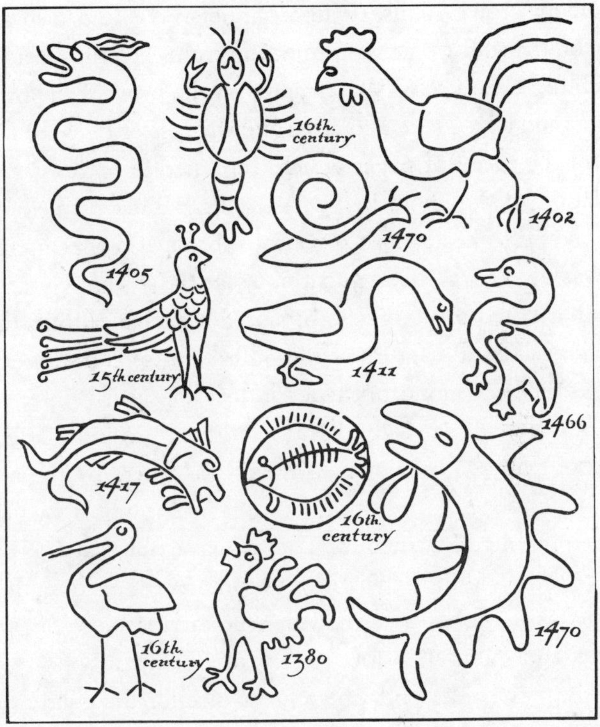

NO. 194

The bull was a favourite watermark with the old papermakers and this animal in hundreds of forms may be seen in books from the fifteenth century onward

as well as snakes, fish, snails, turtles, crabs, scorpions, and all varieties of insects. It was the forms of animals that required most dexterity to twist in wire and which gave the early artisans the greatest outlet for their skill. The bull's head, one of the earliest animal papermarks, made its appearance in 1310 and was a favourite emblem with the papermakers for over two hundred years. The head of the bull was sometimes used without appurtenances, but more often it is found surmounted by a Latin cross, rose of bliss, half-moon, crown, flails, or like symbols. We find also a curled snake on the staff of a cross projecting from the bull's head device, resembling a caduceus. Watermarks of the head of the bull, both surmounted by a snake and cross and the rose of bliss are found in some of the sheets of paper which were used in Myles Coverdale's translation of the Bible in 1536—the first Bible printed in English. The bull's head mark in various forms also appears in paper that was used by William Caxton, Colard Mansion, Gerard Leeu, and other noted printers who procured their paper from the Low Countries. The full figure of the bull is not so common a watermark as the head alone, but when it does appear it is interesting to note that the tail is usually divided into three strands. The bull as a symbol was handed down to the papermakers by the ancients, for chief among the multitude of idols and symbols was the god Apis, represented by the bull. The ox was emblematic of patience and strength, and in the writings of some of the early Fathers it is accepted as a symbol of Christ the true sacrifice, also of prophets, apostles, and saints slain for the sake of Christ, and of all who patiently bear the yoke and labour in silence for the good of others.

In illustration No. 197 may be seen the evolution of the bull's head as a watermark. The first mark in this figure shows the bull's head from the Myles Coverdale Bible of 1536. At the beginning of the sixteenth century the head gradually developed into a shield-like scroll, still retaining the horns on each side at the top, as shown in the second drawing. The third tracing shows the watermark used by William Rittenhouse, thought to be an outgrowth of the original bull mark of the fourteenth century.

The unicorn, like the bull, is found as a watermark in the paper used by Caxton, and holds a prominent place in papermarks from the fifteenth to the seventeenth century. M. Briquet has recorded over eleven hundred examples of this animal used as a device for marking paper. The unicorn was symbolic of purity and inno-

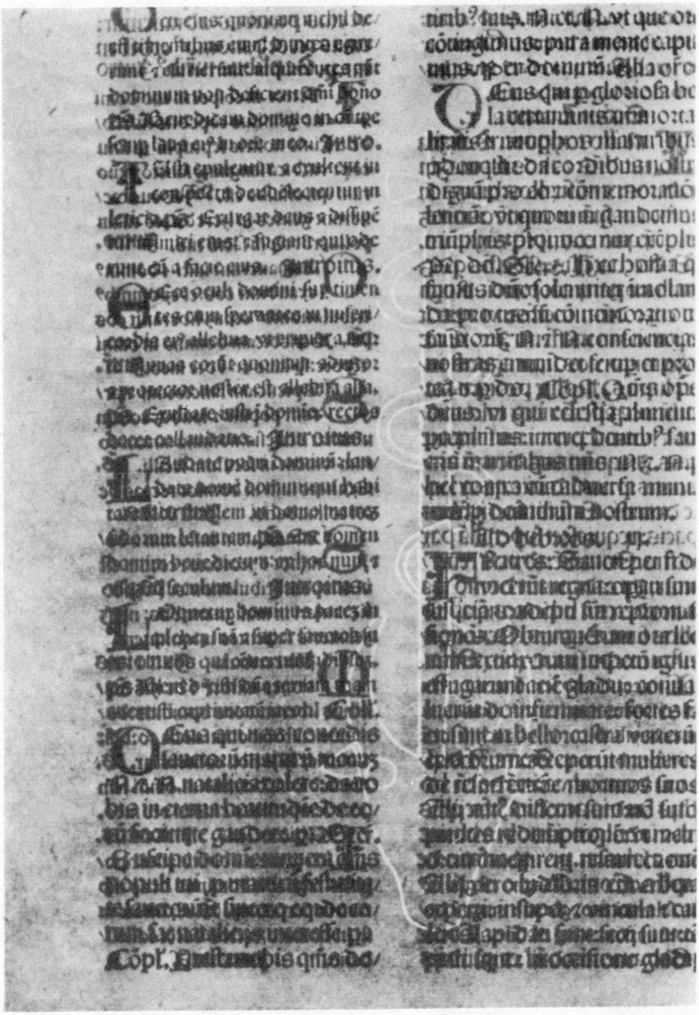

NO. 195

An unusually clear bull and snake watermark. In a copy of *Missale Frisingense* (Hain 11303), printed by Erhart Ratdolt, Augsburg, 1492

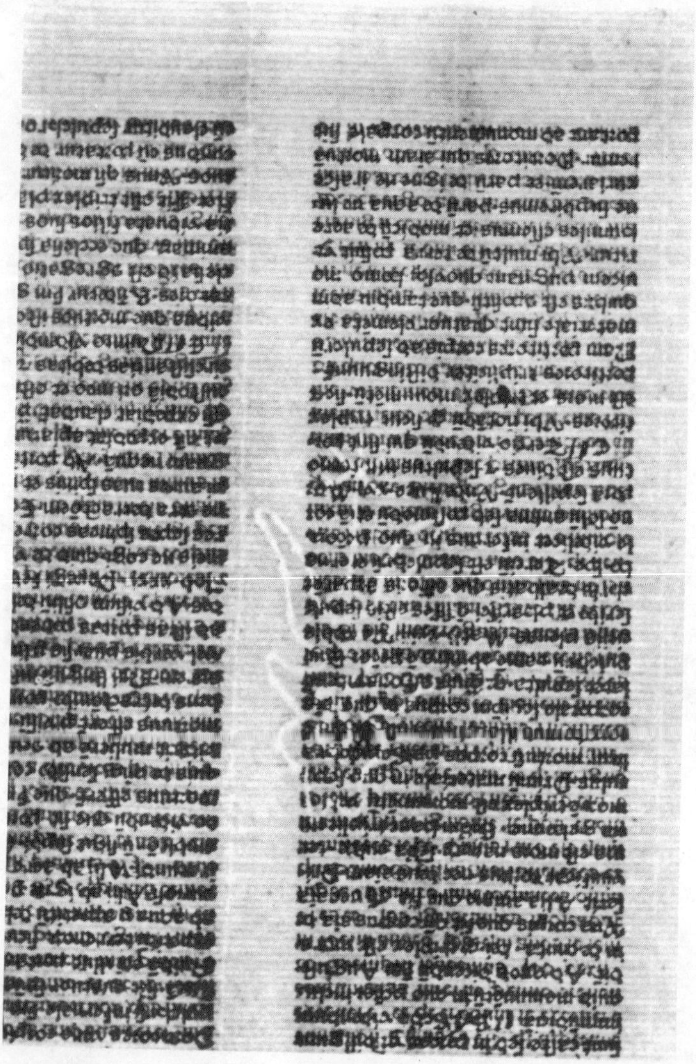

NO. 196

A well-defined bull's head watermark. Found in a copy of *Johannes Nider: Sermones* (Hain 11803), printed by Peter Drach, Speier, 1479

cence, and it was believed that the horn of this mythical animal was a panacea for all illness and an antidote for poisons. The ancients believed the horn of the unicorn so sensitive that if a cup of poison was brought near it a thick moisture would be expelled

NO. 197

The evolution of the bull's head watermark, from 1536 to the mark of William Rittenhouse of 1690

from the surface of the liquid, and if a piece of horn was thrown into the poison, the poison would bubble and in time boil over. It was also thought by the ancients that the horn was movable at will, like a kind of sword. The unicorn was used as a watermark by the old papermakers both in full body and in head alone. It is seldom found as a papermark with any other symbol incorporated with it, but in a few instances a sword or cross protruding from its back has been noted. A mark of this kind is shown in one of the specimens from 1479. The wire forms for both of the watermarks illustrated, from the year 1479 were applied to the paper moulds parallel with the chain wires. This was an uncommon practice, as it was customary for the old workers to place the watermarking wires parallel with the laid wires, or so that the device faced the vatman as he held the mould.

Another favourite animal watermark was the form of the dog, and numerous specimens are found extending over a period of about two hundred years. Like the bull and the unicorn, the dog device was used as a symbolic emblem by the ancients, and cen-

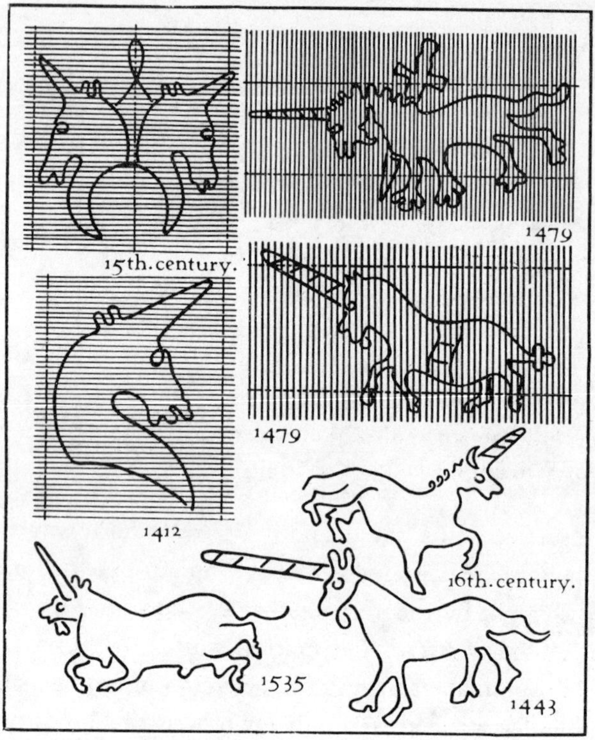

NO. 198

The unicorn, like the bull, held a prominent place as a watermark from the fifteenth to the seventeenth century

turies later was adopted along with other symbols, by the papermakers for their watermarking. Certain species of the canine were considered sacred by the Egyptians and there are instances where sacred dogs were mummified by those ancient people. Early watermarks of the dog usually represent the greyhound, and a great many of them are drawn with considerable motion and vigour, a merit not often found in old papermarks.

The camel is prominent in the watermarks of the fifteenth century and is usually seen in rather a grotesque attitude and of crude

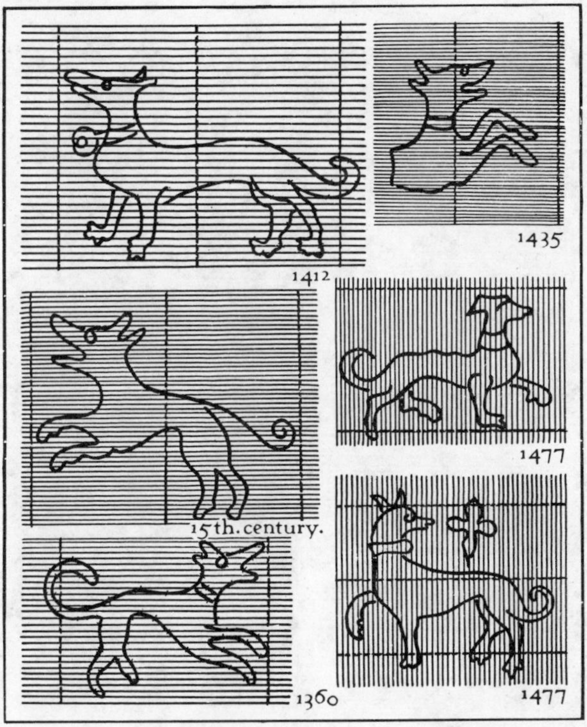

NO. 199

The dog device was used as a symbolic emblem by the ancients, and centuries later was adopted by the papermakers for their watermarking

workmanship. The camel mark here illustrated shows the actual wire form on the surface of the mould. It was constructed of two wires, one starting at the ears and continuing around, shaping the tail and legs and terminating at the neck, while the other wire

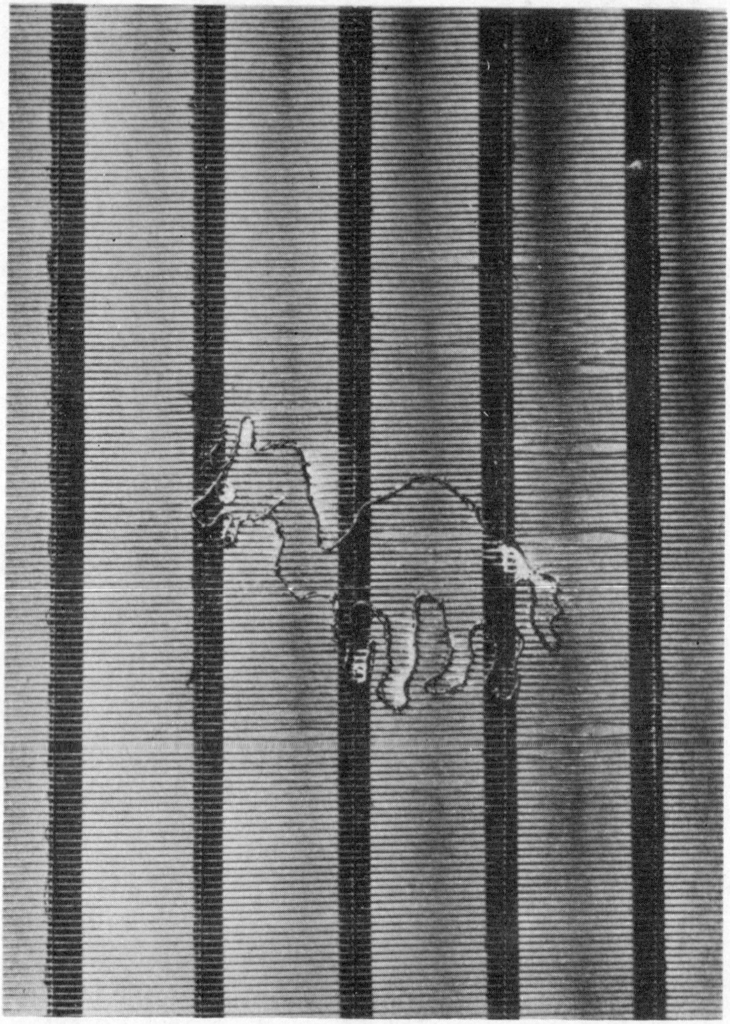

NO. 200

A camel watermark of the fourteenth century. The actual wire form composed of two wires and laced to the surface of an antique laid mould

NO. 201

Animal watermarks of every variety are found in papers from the middle of the fourteenth century onward

completed the head, the eye being shaped by a crook in the second wire. This camel papermark is French, from about 1379.

Elephants, leopards, goats, lambs, dragons, cats, horses, and deer are found in abundance in watermarks from the middle of the fourteenth century onward. The cock is not a common mark in old paper, but a variety of specimens, mostly of French origin, have been discovered in the pages of antique books and manuscripts. Many of the chanticleer marks display ability in design as well as in fashioning in wire. It is noticeable that the cock was generally formed with open bill, which, according to the authorities on symbolism, denotes the dawn of light. Birds of many kinds and sizes furnished a multitude of devices for the old paper workers in marking their handiwork. We find also in unlimited numbers, representations of fish, as well as of every crustacean animal of the sea.

During the first fifty years of papermaking in the American Colonies there still existed among craftsmen a love for symbolic design which was manifested in the watermarks of the period. But by the middle of the eighteenth century superstition and symbolism began to lose hold upon the artisans, and from that time, in the Colonies as in Europe, symbolic watermarks and printers' marks began to fall into disuse. Watermarks with an emblematic significance were not uncommon in the early papermaking history of Pennsylvania, and the pioneer workers of this region were responsible for the most unusual specimens. New England, which contributed so much to the history of the applied arts in other fields, furnished little of an artistic or interesting nature to the watermark annals of America. The few marks that are recorded from this locality consist chiefly of names and initials of the papermakers, which have no special appeal to the imagination.

The first paper mill in the Colonies was established in Pennsylvania by William Rittenhouse, a native of Germany, who was assisted in his undertaking by William Bradford, the first printer in Pennsylvania and New York, and two other worthy gentlemen. In all probability the papermaking moulds used by this mill had been brought from the principal founder's native land and were no doubt plain moulds, without watermarks. The first watermark used in Rittenhouse paper was the single word "Company," designating the original partnership. Judging from its crudity it was fashioned by untrained and unskilled hands, probably by one of the mill workmen unaccustomed to forming objects in wire. This mark was used from about 1690, when the mill was established, until sometime during the year 1704, when Bradford was induced to part with his share in the paper mill. In 1706 the property was in full possession of Rittenhouse. The second papermark to be

NO. 202

The second watermark adopted by the Rittenhouse paper mill

adopted by this pioneer mill was the monogram WR, the initials of the principal founder, on one half of the sheet of paper, while on the other half appeared a clover-leaf inside a shield which was surmounted by a crown. Underneath the shield, in outline letters, was the word "Pensilvania." The shield and crown device displays a great deal of Dutch and French influence, as watermarks of this style had been used by early papermakers in Holland, adopted by them from France where the device had been used as a papermark as early as 1460. This emblem was an evolution of the bull's head watermark as previously explained. The crown and shield with three fleurs-de-lis constituted the arms of France, and this device was frequently used as a watermark by the papermakers of the Low Countries, probably in reference to the direct descent of the House of Burgundy from the Kings of France. Papermarks of this character may be found in the paper used by William Caxton, the first printer in England, whose paper was procured from the Low Countries. The clover-leaf, trefoil, or "klee-blatt," which Rittenhouse substituted for the fleur-de-lis, was adapted from the town-mark or seal of the village of Germantown, a settlement not far distant from the location of the paper mill. This seal or emblem had been in use by the community only a short time before Rittenhouse put it to use as a watermark.

The next papermark to be adopted by this pioneer mill of Penn-

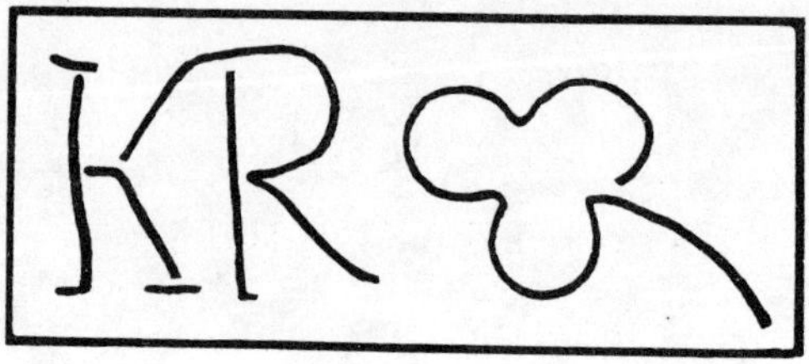

NO. 203
The Klaas Rittenhouse watermark

sylvania consisted simply of the letters KR, the initials of Claus Rittenhouse, the son of the founder, the name in Dutch being Klaas. With the use of this initial mark on the left half, the clover-leaf was sometimes introduced on the right half of the paper. The initials and clover-leaf here reproduced were traced from the paper upon which was printed the first issue of *The American Mercury*. This newspaper was established in Philadelphia the 22nd of December, 1719, by Andrew Bradford, the son of William Bradford, one of the owners of the original Rittenhouse paper mill. The only other watermark known to have been used by this family of papermakers was the letters IR, the initials of Jacob Rittenhouse, a great-grandson of the founder of the mill.

In 1710 William De Wees established the second paper mill in the Colonies. This mill was really an outgrowth of the original Rittenhouse establishment. Apparently the De Wees owners did not make use of watermarks. It is, however, recorded that this mill manufactured "an imitation of asses-skin paper which was well executed."

In the year 1729 Thomas Willcox set up the third paper mill in the colony of Pennsylvania. This establishment used as a standard watermark a dove and olive branch design with the initials of the papermaker—first, TMW for the founder, and in later years the

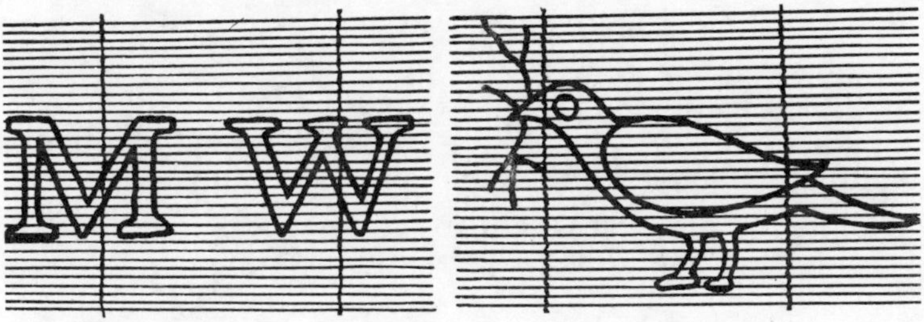

NO. 204
The dove and branch watermark of the Willcox Mill

letters MW, the initials of Mark Willcox, the son, who was operating the mill in 1767. In 1827 an ivy leaf was adopted as a watermark, the mill being known as Ivy Mills because of the English ivy which covered the stone buildings. The original vine was brought from England by Thomas Willcox to Pennsylvania in 1718, from near the Old Ivy Bridge in Devonshire. This mill also used the familiar post-horn device as a watermark previous to 1787. There were a great many bank note papers executed at Ivy Mills, most of which were watermarked with the names and marks of states and banks. One of the earliest of these papermarks was made for Pennsylvania in the year 1777, shown in No. 206. A letter of Revolutionary interest (dated March 11th, 1778), concerning this watermark was written to Colonel Andrew Boyd, sub-Lieutenant of Chester County, Pennsylvania, where the Willcox paper mill was located. This letter reads in part: "Mr. Willcox has in his possession a mould for making paper belonging to this state, which you are requested to bring away, it is marked with the word "Pensylvania" in twenty-four places, he did promise if the

NO. 205
The Willcox ivy leaf watermark

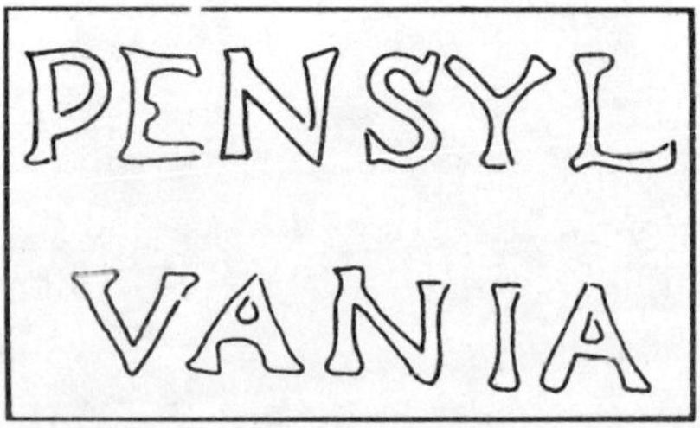

NO. 206
Watermark of one of the earliest bank note papers executed by the Willcox paper mill

NO. 207
A sheet of watermarked bank note paper made by the Willcox Mill for the Bank of Greece in 1859

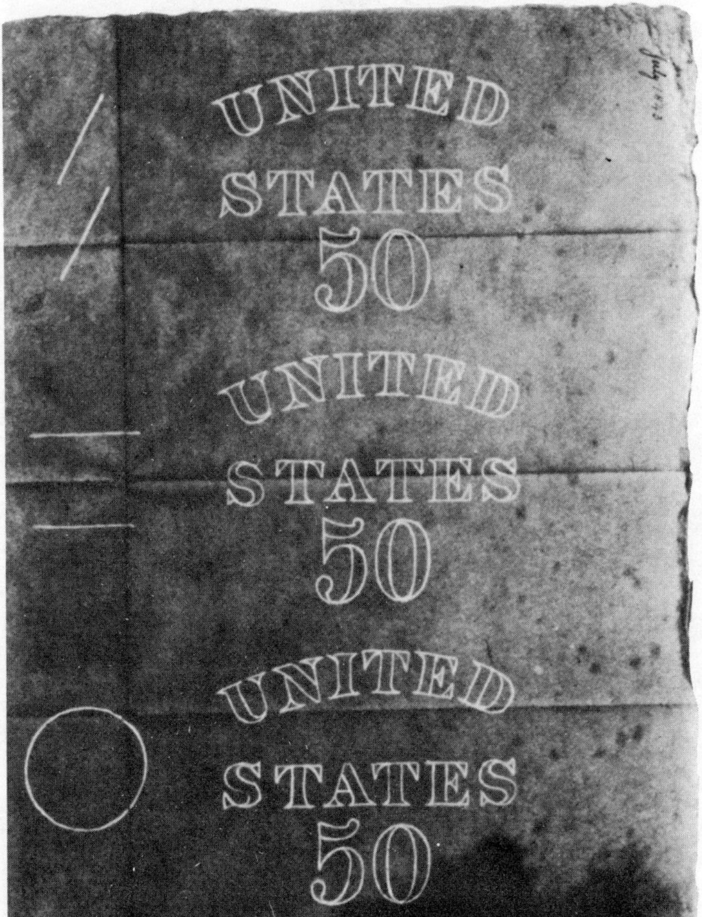

NO. 208

Watermarked paper made for United States treasury notes in 1843 by the Willcox Mill. Only 35,000 of these sheets were fabricated

enemy came that way he would throw it into the mill dam." During the nineteenth century the Willcox mill executed some elaborate bank note watermarks for South American States, the Bank of Greece, Ohio State Bank, and the Bank of Montreal. As

NO. 209

By an act of Congress, approved March 31st, 1848, the President was authorized to borrow a sum not exceeding $16,000,000. The eagle watermarked paper, made by the Willcox Mill, was used for the issuing of these bonds

early as 1838 a finely watermarked paper was made for New Granada, and in 1859 the Willcox mill executed for the Bank of Greece a beautiful mark which consisted of a bivalve shell, a pecten, in addition to the words and figures indicating the denomination of the note. The last handmade bank note paper fabricated at this historic mill bore the letters BM in the watermark and was manufactured for the Bank of Montreal in 1861, one hundred thousand sheets being made.

The Gilpin Mill, also in Pennsylvania, used the dove and branch watermark certainly as early as 1789, possibly from the establishment of this mill two years earlier. This mill in 1793 was using the

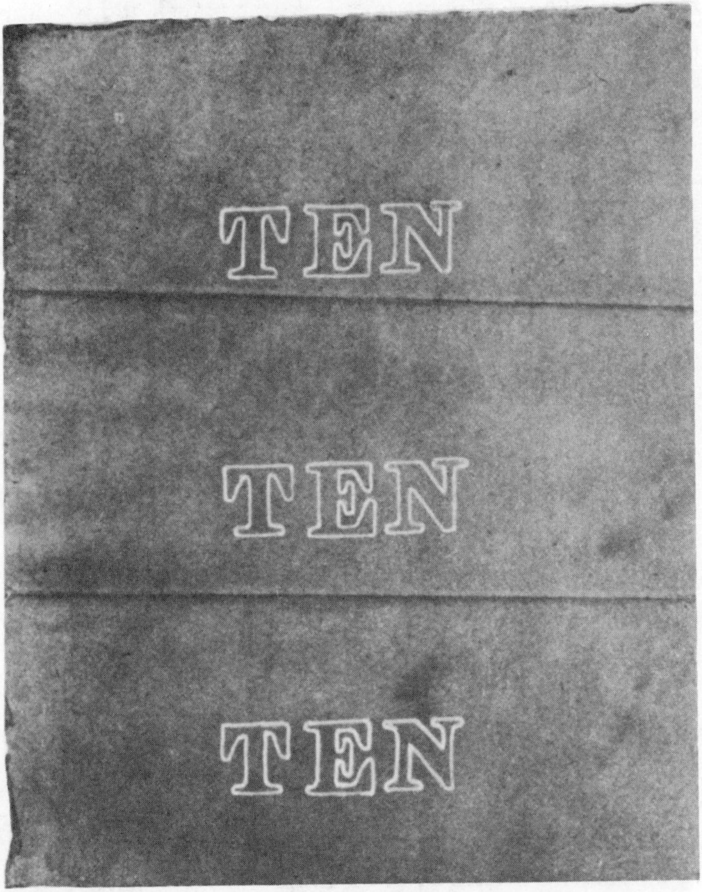

NO. 210
Watermarked bank note paper made in 1850 for the United States Government by James M. Willcox who operated Ivy Mills between 1827 and 1854

word *Brandywine* as a watermark, as the Gilpin mill was situated on the Brandywine River. The dove and branch was a favourite watermark with American papermakers, for we find it used again in 1805 by Thomas Amies, the third mill to adopt this design as a trademark. The dove sometimes lacks the branch and is at times quite poorly drawn, but the identifying name is usually present, variously given as AMIES, AMIES PHILADA, etc. One brand of paper bears the additional mark, CONGRESS U. S. Amies was at one time

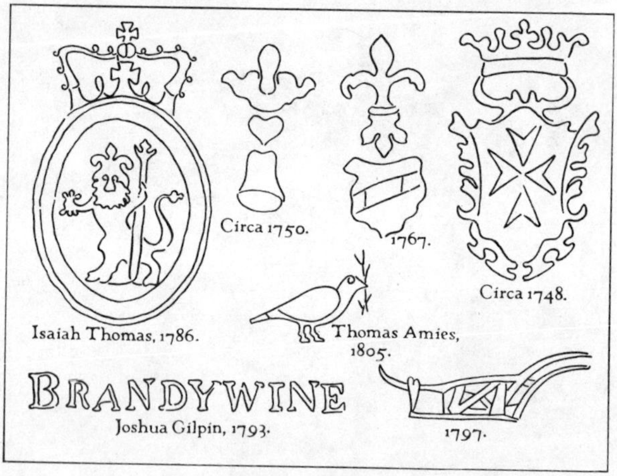

NO. 211

Reproductions of seven watermarks used by Pennsylvania and New England paper mills

superintendent of the Willcox mill at Chester, and after setting up his own establishment he adopted the dove and branch, of Willcox origin, as a watermark. A direct descendant of Thomas Willcox now living in the old family home at Chester has in his possession a pair of moulds that were used in the old Ivy Mills, which bear the dove and branch watermarks.

The New England paper mills which were founded during the middle of the eighteenth century apparently used no watermarks, at least it is not possible to trace definitely many emblems which originated in this locality. In 1786 Isaiah Thomas of Worcester, Massachusetts, was using a figure of a lion enclosed in an oval border and surmounted by a crown, as a watermark on one side of his paper, with the name THOMAS on the other. The watermarks of early New England papermakers are so obscure that I hesitate to make any definite assertions regarding them.

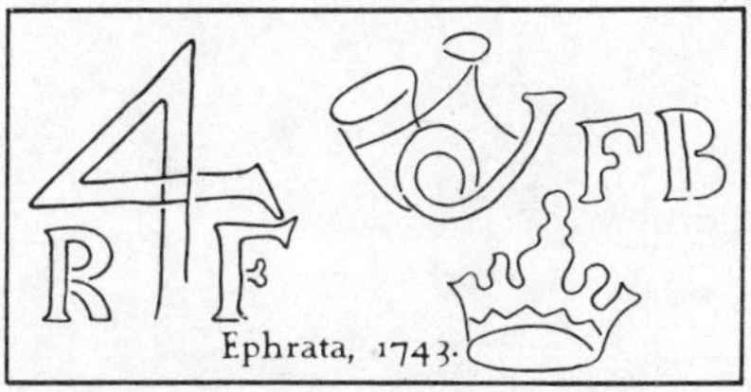

NO. 212
Ephrata watermarks from about 1743

The fourth paper mill in Pennsylvania was established forty-six years after the original Rittenhouse project. This was *Die Papier Mühle des Brüderschaft zu Ephrata,* built on the banks of the Cocalico Creek, in Lancaster County. Ephrata was a communistic settlement, a branch of the Pietists of Germany who emigrated to Pennsylvania in the early eighteenth century. The members of the community lived in a cloister or convent under monastic rules of celibacy and austerity. The community was self-sustaining, as they not only fabricated their own paper, but engaged in printing and bookbinding, as well as in a number of simpler crafts. Several of the buildings of the old cloisters are still standing and form one of the most interesting groups of early eighteenth century construction in this country. The paper mill of the community at Ephrata was under the direction of the Funk brothers, Samuel and Jacob, both experienced in the art of papermaking, having acquired the trade in Germany. The principal product of the mill was a coarse brown paper which they called "macalatur." This paper was never watermarked, but the finer grades of printing and writing papers bore marks of identification. One of the Ephrata papermarks was a large figure four, the perfect number, below which were the initials R F, a private mark of the Funk family. The letters F B for Brüder Funk also appear, without the figure four, in some of the Ephrata publications. This mark may be studied in the paper used

NO. 213
The symbolic watermark of the Ephrata paper mill

in the edition of *Theosophische Lectionen,* printed in 1745. The most emblematic watermark of this idealistic community appears in the paper used in mystic books that they printed. This highly symbolic device consists of a Latin cross surmounted by a scroll with the words ZION, two keys forming triangles with the uprights and arms of a cross, which rested upon a narrow panel bearing the name EFRATA. The keys had reference to the *Clavicula Salomonis,* or Keys of Solomon, a mystic book of the seventeenth century which was highly regarded by the Brotherhood. The entire emblem was capped by a filigree, not unlike the upper part of the second Rittenhouse watermark. The cross and keys device may be found in books printed at Ephrata before 1745.

Several other watermarks which were used by this mill were the crown, the post-horn, and perhaps a three circle device which again shows the perfect figure four, so much favoured by the Brotherhood. In going through one of the dilapidated old buildings at Ephrata many years ago, I found an old chest which had been hidden under a lot of discarded lumber. When I opened this chest I was much surprised to find about seventy sheets of paper, each

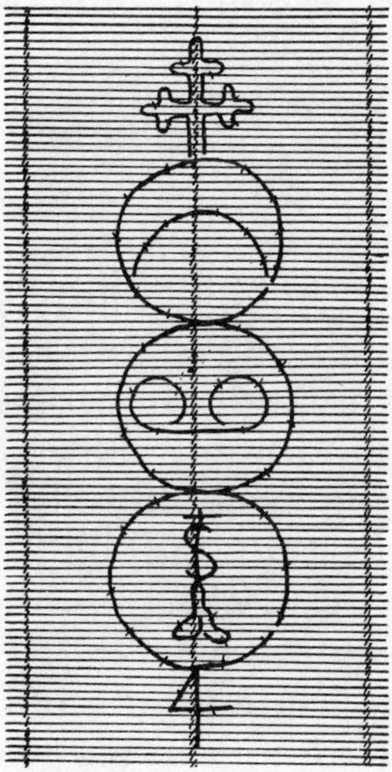

NO. 214

This watermark appears in the old printed sheets found in one of the buildings of the Ephrata Brotherhood

sheet having eight pages of German text printed upon it; the paper was unfolded, just as it had come from the press in the eighteenth century. These sheets are now in my possession, and while I believe the paper to be of European make, it may be that it was formed at the Ephrata mill upon moulds brought from Germany. It is in this paper that the three circle and figure four device appears, but it is not possible to identify it definitely as an Ephrata watermark. The printing, however, is authentic. In the paper upon which early American books and newspapers were printed we find numerous watermarks, but it is an almost hopeless task to classify the marks and state precisely what mills produced them. Papermaking moulds were brought to this country from all parts of Europe and it is probable that many of them had watermarking wires already attached. In this way foreign marks were introduced into American-made sheets of paper.

Each of the watermarks,—whether the simple devices of the thirteenth century, the human figure and head, man's works, vegetation, animals, or the devices of the early American papermakers—appears to have had its place in the world of symbolism, but whether or not they meant anything, other than marks of identification, is conjecture. However, it is reasonable to believe that they bore some significance and held some vital place in the lives of the craftsmen who fashioned them. For students of paper and papermaking it is unfortunate that the early artisans did not see fit to employ as watermarks some object or utensil of their own calling. How interesting it would be to find a mould, a vat, or a paper press pictured as a watermark in a book page from the press of Gutenberg or Caxton; but the old workers were used to seeing and handling their own implements and probably regarded the tools of other crafts and walks of life as more curious and ingenious. Whatever the origin or use of the papermarks, the many thousands of devices which we have inherited from the early papermakers are so varied that they form almost an encyclopaedia of design, and illustrate a multitude of mediaeval objects and utensils that otherwise might not have been recorded. Present-day workers in the paper industry may well be proud of the history and traditions of their trade, for no other craft can show such an endless array of designs extending over a period of almost seven hundred years.

INDEX

INDEX